W9-BBI-353

The Challenge of Command

Reading for Military Excellence

The Challenge
of Command

Reading for Military Excellence

Roger H. Nye

AVERY PUBLISHING GROUP INC.
Wayne, New Jersey

Cover design by Martin Hochberg and Rudy Shur
Cover photo by Martin Hochberg
In-house editor Jacqueline Balla
Typesetting by ACS Graphics Services, Fresh Meadows, NY

Library of Congress Cataloging-in-Publication Data

Nye, Roger H. (Roger Hurless)
    The challenge of command.

    Bibliography: p.
    Includes index.
    1. Command of troops. 2. Command of troops——
Bibliography. 3. Military art and science——Bibliography.
I. Title.
UB210.N94 1986   355.3'3041   85–30614
ISBN 0–89529–280–7 (pbk.)

Copyright © 1986 by Roger H. Nye

All rights reserved. No part of this publication may be reproduced, stored in a
retrieval system, or transmitted, in any form or by any means, electronic,
mechanical, photocopying, recording, or otherwise, without the prior written
permission of the copyright owner.

Printed in the United States of America

10 9 8 7 6 5 4 3

Contents

General Creighton W. Abrams with Lt. Col. Robert C. Bacon, Commander, 3d Bn, 21st Infantry near Tam Ky, Vietnam, 1969.

In Memory of
Creighton W. Abrams, Jr.
who mastered the art of command
and taught it to us by his example
1914–1974

His commands included:

37th Tank Battalion, 4th Armored Division	1943–1944
Combat Command B, 4th Armored Division	1944–1945
63d Tank Battalion, 1st Infantry Division	1949–1951
2d Armored Cavalry	1951–1952
3d Armored Division	1960–1962
V Corps	1963–1964
U.S. Military Assistance Command, Vietnam	1968–1972
Chief of Staff, United States Army	1972–1974

Preface

This book of commentaries about military command was conceived to answer, at least in part, the question that shadowed me through four decades of working with military students: "Sir, what should I read?" There is, of course, no general answer to this very individual question. I have, however, encouraged all to keep asking that question, remembering General George S. Patton who, according to the movie *Patton*, watched the Afrika Korps wither before his defenses and shouted, "Rommel, you magnificent bastard, I read your book!"

When I asked a veteran military reader, Peter M. Dawkins, what lieutenants should read, he suggested that they try to gain a vision of what they might be as future military men and women. Paul F. Gorman, the most inquiring of soldiers even into four-star rank, told me they should read the military history that best addressed problems they might face in their own careers. "Would the generals of World War I have acted differently," he asked, "had they taken to heart the lessons of the machine gun and barbed wire that were so evident a decade earlier in the Russo-Japanese War?" Finally, George S. Patton, who followed his father in command of the 2d Armored Division, told me that they should read accounts of the Great Captains, for guidance on how to command fighting organizations.

And so I was guided initially towards a book for a rather select audience. It is for those inquiring soldiers who continue to ask questions concerning what to read, and who are comfortable with the study of history, with some offshoots into the humanities and social sciences. Furthermore, these commentaries are for those who aspire to command, and who seek a vision of themselves in the many roles inherent in such a complex undertaking.

Very few men and women who enter the Army have the opportunity to command battalions and larger military organizations; there are far more command clients than command billets. Why, then, should all officers focus their reading on command? The obvious answer is that the Army needs a pool of potential commanders in case of a largescale war. An equally important answer is that all officers will perform better in their staff and specialist roles if they see their work through the eyes of the commander, with an understanding of his responsibilities and needs. A third answer is that the common study of command serves as a cohesive force in an Officer Corps that is being fragmented by specialization. The commander must be the generalist. It is in his mind that all the parts of the puzzle are pulled together. All officers need to understand this mind.

Officers begin to learn the art of command from their seniors in their first days as cadets and lieutenants. Most senior officers assume it is their duty to counsel juniors about command, and to advance their knowledge and skills in that art. If these commentaries are of help in organizing officer professional development programs, so much the better. In furtherance of such an object, the text is organized to: (1) give an overview of the command phenomena; (2) analyze the field into manageable topical chapters; (3) suggest major problems and trends in each topic; (4) highlight some of the best books on each topic, based on readability, substance, and availability; (5) raise questions for reflection and discussion; and (6) include a detailed bibliography for ease in locating books that have been given an abbreviated cite in the text.

The writing of these commentaries had been in gestation for nearly a decade when Mr. Rudy Shur, Managing Editor of Avery Publishing Group, suggested that they serve as a guidepost to *The Art of Command Library*, a series designed to republish some of the best military histories, biographies, and monographs that had been unavailable to the military reader in recent years. We debated whether commentaries based on my Army experience would be useful to students of command in the naval and air services, and to young citizens without military experience. The answer was yes, in that all services address command in much the same language in matters of professionalism, the warrior spirit, strategic goals, management, and responsibilities to the American public. In matters of tactics, training, military hardware, and doctrine, however, a commentary that suits the Army and the Marines needs to be augmented for students of Navy and Air Force command, due to the uniqueness of the missions and the technology of these services.

In this book, we have tried to minimize the use of Army in-house jargon that might confuse the layman or cadet who has yet to undergo extensive military experience. This is not always a successful venture, for military people glory in their acronyms and computerese. In that vein, many of the current issues addressed in these commentaries were generated in the teleconferencing networks of the U.S. Army Forum, in which informed and creative soldiers and civilians freely debate proposals for new directions for developing leaders, trainers, and commanders in Army programs.

Women are increasingly assuming the tasks of military command in units not destined for combat duty. This is a wanted change. My writing, however, is drawn from the books of a 2000-year heritage, in which military commanders were male. I cannot distort this historical record with "his or her" when the authors did not intend so. Until a new literature is written, I trust the reader to infer both genders where that can be correctly construed for the future.

My deep thanks go to the soldiers who carried the other end of the dialogue in these commentaries, and to the many others whose voices were always in my head. While they may not endorse many of the conclusions I came to in this undertaking, they kept me focused on the real world that the commander inhabits, and they required the theoretical literature to conform to that reality. If there is usefulness in this book, it is because of that insistence on their part.

Roger H. Nye

The Rocks, Webb Lane
Highland Falls, N.Y. 10928

CHAPTER ONE

Visions of Our Military Selves

By force of will and against his inner disposition, he created himself in the image to which he aspired.

—Blumenson, *The Patton Papers*, 7.

Inquiring soldiers, by definition, are determined to expand their universe through reading. But, to what ends should they direct this energy? Some soldiers would say, with utter frankness, that they hope to improve their chances for high command and greater glory. They should proceed with caution, for the history of warfare is studded with powerful men who did not let a book encumber their march to a pedestal in Valhalla. Reading is related more to the essence of success than to the trappings of it.

Closer to reality is the view of military historian Theodore Ropp of Duke University, who suggests that the most durable reason for reading military history is simply "because it's fun." That established, he might go on to suggest that the professional can also parlay this fun into a better understanding of how to build fighting units and to nourish one's technical and tactical competence—subjects to be discussed more fully in the following chapters.

Many inquring soldiers, however, are still going through those novice years of cadetship and on-the-job subalternship, when one's commitment to professionalism may not have blossomed, and reading has not yet become "fun." To what ends does the soldier read when he or she is part military, part civilian, and always on the run?

Answers to that question lie more in the troop units than in academia, in the minds of commanders like Colonel Peter M. Dawkins. When I found him in 1979 at Fort Campbell, Kentucky, he had devoted the past five years to leading military organizations designed to train and fight. On and off for two decades, he and I had debated the question of military learning, based on his academic pursuits at West Point, Oxford, and Princeton. Now he was commanding the 3rd Brigade of the 101st Airborne Division (Air Assault), and was busy with helicopters, light infantry tactics, Strike Force exercises—and some 75 lieutenants who populated the three battalions under his command. When I asked him what his young officers needed to learn, he answered as follows.

"Most lieutenants who have been in a company for a year know their jobs reasonably well. They've picked up the fundamentals as cadets, and in the Officer Basic Course at their branch school. But real learning *for* the job comes from being *in* the job. When the fresh, recently commissioned lieutenant reports for duty, it becomes immediately clear to him what he must learn—and the failure to rapidly develop the requisite competence frequently results in moments of terrible embarrassment. So he learns from others, from manuals, from making mistakes.

"Unfortunately, that's where it tends to stop—just this side of what they need to know each morning to get today's job done. There is little learning about that future which lurks just around the corner, when they may find themselves captains in combat, or in command of combined arms teams with artillery and air support to deal with as well."

Dawkins continued: "The focus of 'lieutenant learning' is properly on what an officer *does*. But, more needs to be said about what an officer *is* and *should be*. Very few junior officers have any clear sense of themselves as military men with responsibilities and opportunities that are unique in American life. They seem to plod along day to day, working hard and developing the essential skills of their profession. But many—if not most—miss a critical step: that of developing a sense of how their growing confidence and competence relate to the essence of what a soldier is meant to be.

"I don't believe it's naive or unrealistic to expect junior officers to read. It wouldn't take all that much time or effort, and doing so would help them gain a much richer and fuller image of themselves as soldiers. What they need is a template, a guide, a vision of what being a soldier today is all about, and what it may come to be in the future.''

Dawkins' observations were well supported by other sensitive commanders, and by the psychologists who have written extensively in motivational theory about the fulfillent of aspirations. But how does one read to gain a vision or visions of himself? By reading biographies for the visions that others have had of themselves? By reading Shakespeare for the visions of fictitious heroes? By reading history? Psychological theory? Yes, all, probably.

But does not each person have a different personality, a unique chain whose links of genes, experiences, and chances are like no one else's? Is it not, then, that each soldier must find a particular vision of the future for himself, and that each one will find something different in a given book? What is inspirational for some may be very tiresome for others. A reading list for all may be useful reading for none.

THE MANY USES OF OUR VISIONS

It has been said that a vision is an acute sense of the possible. My generation of soldiers was the post-World War II generation, and the young Armor officers among us thought it was possible to become future George S. Pattons, Erwin Rommels, or, for those who knew him, Colonel Creighton Abrams. We rode about the training grounds in our tanks and armored cars, with cigars and yellow scarves and tanker boots with straps—boots that would not let us walk more than a few miles. We rode the last of the Army's horses, talked about 25-mile-an-hour warfare, and snickered when the Infantry S-4 sent us gasoline in five-gallon cans to refuel our M4A3E8 tanks. We carried copies of Guderian's *Panzer Leader* and Desmond Young's *Rommel* and Patton's *War As I Knew It*. Fighting in the Korean mountains dampened our vision a bit, and, for many, the Vietnam War was another diversion from the military reality that lay on the European plains where Soviet tank divisions were massed.

And what did our visions do for us? In many cases, they kept us in the military service for a full career. For many, they gave our lives more meaning and lifted us beyond our day-to-day jobs. They put many of us on a quest for new knowledge, helped us grow, and made us more adept at creating new things. They gave us inspiration to reach for excellence, to lift ourselves into professional status, and to earn respect from others. They gave us style.

A few of us were made better able to cope with the great changes that were sweeping through our society during our lifetimes, for we were challenged to make our visions work. Otherwise, we must face the prospect of abandoning them and leaving ourselves rudderless.

Then, too, some of us died for our visions.

It can be argued that the American method of introducing young men and women to military service is not conducive to the formulation of such visions. Attracting qualified youth to the Officer Corps of the armed services is achieved through a contract that offers financial support for college education in return for training in military skills and a commitment to several years of service. If the Army commissions, say, 10,000 new officers in a given peacetime year, some 7000 might attain their commissions through the Reserve Officer Training Corps detachments on more than 400 college campuses. Another 1000 would be commissioned from the U.S. Military Academy, and the remainder would have attended Officer Candidate Schools or received direct appointments as specialists. All but the latter have undertaken their education in the arts and sciences simultaneously with their military training. This is admirable for recruiting and for preparing young people for the real world, which is neither totally Athenian nor totally Spartan. But, this process of combining education and training does not nourish a clear vision of a life dedicated to military service. This dilemma is examined fully by John P. Lovell in his 1977 *Neither Athens Nor Sparta.*

In addition, many have been handicapped by a contemporary educational culture that puts a premium on solving problems at hand rather than constructing a viable philosophy of life. Behaviorists might say that they are task oriented rather than personality oriented. Many are the products of a television diet that narrows one's attention span to the ten minutes between commercials—a diet that accustoms one to thinking in terms of the present tense rather than the past or future, and portrays an unreal world in which problems are both presented and solved in one sitting.

Once commissioned, most junior officers are further burdened in the Army workplace, where life consists of getting up in the morning, going to work, solving problems, going home to bed, getting up, going to work, and so on. In view of these limitations on learning and outlook, many can expect an intellectual stagnation to set in by the age of thirty.

With it comes the atrophy of their ability to remain excited about their work, to create new answers to old questions that never get solved, and to adapt to change. Building a vision of their future selves is a powerful antidote against these outcomes.

Where does the inquiring soldier find that broad array of written visions that has inspired military men since Joshua? When I ask my colleagues, the historians, they are sure to start chattering about military biography, that treasurehouse of stories about Alexander the Great and Caesar and Napoleon and Wellington and Grant and Lee and Pershing. Why biography? To learn how a soldier thinks and acts, to discover that today's problems have been thought about before, to explore the curious relationship between the soldier and the public he serves, to appreciate the essentiality of courage, to see how others have led their troops into dramatic successes with the barest of resources. Why? For visions of excellence.

But are not the technology and the social system of the time of Caesar or Napoleon so far removed today that the reading exercise is of little real use to the contemporary soldier? And is it not unwise to counsel that the soldier-tyrants of ancient days have lessons for the soldiers of a democracy? Are not Mao and Giap and Castro better heirs to the legacy of the politico-military generals of the past? We historians argue that the worthy soldier should know of the wide variety of military styles that man has experienced throughout the ages, so he might know better what *not* to be as well as what *to* be.

Soldiers from minority groups are now better able to build a vision of their role in a military society that for centuries excluded them from full participation. The history of the struggles of black officers to gain the responsibilities and opportunities that can provide a viable military lifestyle is now being written. The fight for the removal of injustices and prejudices is best described by Morris J. MacGregor, Jr. in *Integration of the Armed Forces, 1940–1965*. Female officers, on the other hand, are still restricted from combat service, with its attendant impact on professional careers. A beginning understanding of their plight can be derived from Helen Rogan's *Mixed Company: Women in the Modern Army*.

THE VARIETIES OF MILITARY VISIONS

What, then, are the varieties of visions that have illuminated the lives of military men in our 2000-year heritage? We have already touched on the image of the *Battle Warrior*, that charismatic creature who is accorded a hero's status for his competence and daring. For many, Patton was the quintessence of this romantic vision, although others would insist on Rommel or Montgomery as equally suitable choices. This image of the romantic warrior stands out more clearly as we go back in time, to Kitchener and "Chinese" Gordon of the British Empire; to the nineteenth century Americans, "Stonewall" Jackson and William Tecumseh Sherman; and to the still earlier Wellington and Marlborough.

Patton had a vision of himself and he provided a vision for others. In *The Patton Papers*, Martin Blumenson describes how he nourished a conception of himself as a leader of slashing mobile columns in future warfare. By 1943 he was the only man in the American Army who could drive the Third Army across Europe with the strange idea of using an Air Force for his right flank. Through the twenties and thirties he filled diaries and notebooks with close detail, which ranged from Christie suspension systems for tanks to J.F.C. Fuller's new conceptions of mobile warfare. He lectured often in the service schools and reviewed books for *The Cavalry Journal*, including B.H. Liddell Hart's *The Remaking of Modern Armies.*

Patton was probably more versed in military history than any commander of his generation. In one 1928 lecture in Hawaii, he referred to Alexander at Tyre, to fifteenth century paintings in the British Museum, and to the possibilities of splitting the atom. (Blumenson, *The Patton Papers*, 835.) When he was at home in Green Meadows, north of Boston, the house went silent as he drew up his reading chair in the evening. His library, still intact as of 1984, has his annotated copy of Swinton's *Study of War*, a biography of Sir Frederick Maurice, and his Koran, which he studied on the way to North Africa. Blumenson even found mention of Patton's reading in the medical records:

> Patton's professional study during his years in the Office of the Chief of Cavalry was enormous. He now wore glasses when reading. In the spring of 1930 when he suffered . . . an inflammation of the eyes, the doctor recorded that the probable cause was Patton's staying up with his books until one o'clock every morning.
>
> —Blumenson, 870.

Patton's vision of himself as a Battle Warrior was carefully nurtured. The reading and experience of other soldiers took them in quite different directions. George Washington fought as a warrior, but is remembered more as the consummate *Soldier Statesman*, who used his military knowledge and skills to give shape to a great national movement, the creation of the United States of America. The modern-day American Soldier Statesmen, who have had equally decisive impact on the shape of domestic and international events, must include John J. Pershing, George C. Marshall, and Dwight D. Eisenhower.

"Ike" and Patton shared much the same vision of the future when they worked together in tanks at Camp Meade, Maryland, after World War I. But Eisenhower obeyed the orders of the Chief of Infantry to desist from publicizing a mechanized Army that no one could afford. Subsequently, he served in a variety of assignments that equipped him to become a leader of allied armies, such as duty in France in the late twenties with the Battle Monuments Commission, a stint in the Office of the Secretary of War, and the long years with MacArthur in the Philippines. In 1937, he told his family that he had had a fine career and expected to retire as a lieutenant colonel. But, by that time, he had the background and had done the necessary reading to assume staggering responsibilities in World War II.

Eisenhower's background included being first in his class at the Command and General Staff College, which stemmed from long hours of hard work and an unusually broad reading program that he had started in Panama in the early twenties. His Brigade Commander, Brigadier General Fox Conner, had lent him some military novels, of which he later wrote:

> These were stirring stories and I liked them. . . . and so I read Grant's and Sherman's memoirs, and a good deal of John Codman Ropes on the Civil War. I read Clausewitz's *On War* three times. . . . Fremantle's account of the Battle of Gettysburg, as well as that of Haskell.

> It is clear now that life with General Conner was a sort of graduate school in military affairs and the humanities, leavened by the comments and discourses of a man who was experienced in his knowledge of men and their conduct.

> —Eisenhower, *At Ease*, 183–185.

In reading biographies of both Battle Warriors and Soldier Statesmen, the inquiring soldier becomes aware that each generation has a unique demand put upon it, which is to provide a kind of senior military leadership that meets the desperate needs of the time. A good biography explains these needs and how the soldier's youthful experiences prepared him to meet these tests of wisdom and action in later years. Particularly important is the test of combat, which nearly every Great Captain traversed early in his career. The Mexican War was as important to the development of a Lee or a Grant as was the Philippine experience to Pershing and Marshall, and the World War I experience to Patton.

Some military men have been stirred by a vision of themselves as "The Brain of the Army," as *Thinkers and Deciders*. They see their contributions to posterity as a high batting-average in being right in their many decisions which direct the energies of large numbers of troops. Such men are not trapped into the thinker-versus-doer syndrome that has provided the lazy and the arrogant with an excuse for mental mediocrity. Rather, they accept that the cerebral process of making sound judgments is at the heart of military competency, and they study and practice the thinking and decision-making process in a conscientious manner. Sir William Francis Butler had such commanders in mind when he wrote in *Charles George Gordon* (London: 1907, 85):

> The nation that will insist on drawing a broad line of demarcation between the fighting man and the thinking man is liable to find its fighting done by fools and its thinking done by cowards.

Part of the vision of Thinkers and Deciders comes from the better narrative histories of American wars. In *The War to End All Wars,* Edward M. Coffman, for example, tells nicely the story of the American experience in World War I. In so doing, he probes the lives of professional soldiers who never saw Corps or Armies before they were sent to Europe with 2 million American soldiers, and were required to build and fight these massive organizations, with war-weary European generals hoping to siphon off these fresh troops to shore up their own units. This called for creative decision-making in totally new environments, from men who had been spawned in the Indian and Spanish-American wars of another era.

A good military biography illuminates how the soldier arrived at his most significant decisions—how well he defined the problem, gathered his information, formulated alternative solutions, consulted experts, and phrased his decisions. The best biographers focus on the possibilities in a given dilemma, try to assess praise and blame for how the dilemma was handled, and attempt to relate the soldier's resolution of it with his educational background, experience, and moral fiber. In the great modern biographical novel, *The Killer Angels,* Michael Shaara provides a superb analysis of the processes used by the American generals at Gettysburg in thinking about and ultimately deciding on the fate of thousands. For the fate of a lieutenant thinking out a problem in local defense, no one has equalled E. D. Swinton's playful sketch, *The Defense of Duffer's Drift.*

Patton was so aware of the importance of his thought processes that he added a chapter to *War As I Knew It,* entitled "Earning My Pay," wherein he related a series of vignettes in which he made decisions of critical importance. In some cases he decided on more boldness than others dared, and in other cases, more caution. In one case he opted to lie. These decisions led him to observe that a man's military life is lived for about three minutes, parcelled out in bits and pieces of decision making that he alone can make correctly, because of his unique experience and learning.

The image of the military Thinker and Decider has become so complex that the content of the thought process has been analyzed into component parts: tactics, strategy, logistics, administration, research and development, management, and so on. Military men become known for their competence in one or several of these fields, a competence that is nourished early in a career by professional reading. At about the same time that Eisenhower was rereading Clausewitz's strategic conceptions in the 1920's, Patton, the budding mobile warfare genius, was writing to his father, "I have read two books and written some notes on the defense of columns against Air Planes." (Blumenson, I, 808) The inquiring soldier can best begin the search for his own vision as Thinker and Decider with the memoirs of military men who write of their own trials in making critical choices.

One of the enduring visions of the modern military man has been that of the *Officer and Gentleman,* that peculiar blend of authority and morality that connoted competence, good manners, restraint, and elitism. The origin of this conception is European, best seen in Elizabeth Longford's biography of Wellington. She writes that, when he was commanding in India in the early 1800's, Wellington would open his officers' calls with the statement that he expected them to be gentlemen and officers, "and in that order." He would then point out that gentlemen do not accept bribes from Indian princes, nor do they falsify reports, nor indulge in whatever practices he felt obliged to condemn on that particular day. (Longford, *Wellington: The Years of the Sword,* 99) Officership was seen as a technical and legal position that one assumed in a military hierarchy. How one conducted himself in matters of honor, courage, and civility was dictated by his background and education as a gentleman.

In his book *The Image of the Army Officer in America,* Robert Kemble describes how the concept of the gentleman was brought to the colonies and fused into the institutions and laws of the new nation. He suggests that the concept of military gentleman took many forms in our early national years—from George Washington's orientation as a patrician landowner, to Thomas Jefferson's vision of the military gentleman as scientist, engineer, and explorer of the West. Kemble describes the waning concept of the gentleman, after the Civil War had pressed upon the American mind the necessity for looking upon the modern soldier as a professional. The soldier now required a special education if he was to be a practitioner of modern military science, and he would be advanced on the basis of merit rather than class origins. Robert E. Lee became the new model, exemplifying the gentleman professional in a blend of patrician background and a lifetime of study and practice in the new military technology.

THE PROFESSIONAL VISION

Professor Samuel Huntington, in his 1956 *The Soldier and The State,* traced the concept of the military *Professional* through two World Wars. More importantly, he provided the first thorough analysis of the nature and scope of professional officership. In 1960, Morris Janowitz wrote a "social and political portrait" of American officers in *The Professional Soldier,* and suggested new requirements for professionals in the future. He amplified this view in a book of research studies, *The New Military,* in 1964. In 1975, Sam C. Sarkesian added further to the analysis in *The Professional Soldier in a Changing Society.* In this literature there is enough fodder to provide inquiring cadets and lieutenants with an image of what they might be as military professionals through a career of service. Inquiring soldiers, however, may also come away with the question of whether the professional ideal is beyond their reach. A close reading of this literature reveals a staggering challenge to the will and intellect of aspirants. Chart 1 provides a partial summation of these rigors.

Chart 1

The Social Scientists' Criteria
for the Ideal Professional Soldier

Samuel Huntington (*The Soldier and the State,* 11–17) suggested in 1956 that the test for one's fitness to be a professional soldier lies in one's *willingness* to:

- take responsibility for the military security of the state
- become a manager of violence, which includes organizing, equipping, training, planning for, and directing operations of a military force
- treat this as an extraordinarily complex intellectual skill requiring comprehensive study and training
- be motivated by the technical love of craft and the sense of social obligation to use the craft for the benefit of society
- customarily live and work separated from that society
- adhere to the profession's corporate needs by giving support to its systems, which provide for appointment to rank, assignment to offices, schooling and professional certification, and maintenance of professional associations, customs, journals, traditions, and ethical codes

Morris Janowitz (*The Professional Soldier,* 128, 217) required in 1960 a *willingness* to also:

- believe that a military career is materially worthwhile, personally rewarding, and socially useful
- believe in the importance of career success through continuous hard work and self education
- live by the four tenets of professional honor—gentlemanly conduct, personal fealty, loyalty to a self-regulating brotherhood, and pursuit of glory

Sam Sarkesian (*The Professional Soldier in a Changing Society,* 7–13), further required in 1975 a *willingness* to:

- subordinate family, friends, and status in life to professional interests
- accept the public life of living in a goldfish bowl, open to scrutiny as the profession requires
- be regulated by the profession . . . and be a part of the punishment system for those who do not live up to these standards

---

If the novice accepts the social scientists' criteria for professionalism as a statement of ideals, never to be achieved fully in any human life, he can find a useful vision of himself in the professional literature.

That vision entails acceptance of the criteria as a guideline for a living philosophy. Such acceptance is a moral and intellectual problem of no small dimension for men who have had little real world experience in the military. But a consideration of these criteria is essential if one is to adopt the professional model as part of a self-image.

Several questions are inescapable. First, can all soldiers be professionals? Huntington says no, and he rather unsympathetically rules out all enlisted men because they *apply* rather than *manage* violence. Nor do they need the education that is necessary for professional judgment at the higher levels of responsibility. Many veterans would dispute this assertion, on the evidence of the superb noncommissioned officers they have seen "managing" violence in a thoughtful way.

Second, can a reservist become a professional? Yes, says Huntington, some part-time officers can, but they are not able to reach the zenith of skills and knowledge that can be obtained by full-time soldiers.

Third, when is professionalism achievable? Decades ago we thought that a good man had to experience years of cadetship and junior officer duties before he acquired enough knowledge to be a certified professional. We estimated this to be seven or eight years after reporting for duty. Sarkesian, however, feels that the real test is personal commitment to professional values, and that this is proven when an officer voluntarily extends his active duty beyond the service obligation he undertook upon commissioning. Some might make this commitment with as little as three years of service. In short, there seems to be little agreement on the criteria for deciding when one ceases to be a novice and becomes a professional. All can agree, however, on the essentiality of committing one's self to live by professional ideals.

The writers on military professionalism raise a fourth question, which stems from their fascination with the attitudes of the American public towards the military. They seem convinced that military men and women are so concerned about their social images that they judge their own worth on whether Americans have respect for military professionals or consider them dangerous, immoral, slothful, or worse. The military sociologists have a point, of course; no one likes to be categorized among the lesser elements of human society. But so many problems adhere to the analysis that it is not surprising for so many soldiers to seem utterly indifferent to the positive and negative attitudes reflected by the opinion polls.

Kemble found a fragmented public view, pointing out that pacifists have a reflexive, low regard for warriors, whereas those who would "smite the enemy before he smites thee" have a naturally high regard for military things. This has the merit off accuracy, but does not help us get to the heart of overall public opinion. In 1960 Janowitz wrote: "In the United States the military profession does not carry great prestige. . . . Officership remains a relatively low-status profession." (*The Professional Soldier,* 3) On the other hand, Sarkesian begins his 1975 book: "Until the involvement in Vietnam, the American military establishment enjoyed a relatively exalted position in the eyes of most Americans. Rooted in the past, the image of the military man combined the virtues of the heroic leader with those of the unselfish warrior and devoted servant." (Preface and ix)

This apparent contradiction is compounded when Sarkesian continues:

> Since the Vietnam War, however, the military has become associated with the evils of that war, symbolic of the many wrongs and ills of the society—growth of the military-industrial complex; the exposure of graft, corruption, and lack of integrity within the circles of the military; the My Lai massacre; and a general suspicion regarding the moral and ethical behavior of civil servants and public officials—which have done much to tarnish the military image.

Within a few years of this writing, a University of Michigan opinion poll rated military officers among the most respected groups in the nation, above even bankers, politicians, journalists, and Congressmen. In 1983, the University of Chicago National Opinion Research Center found that Americans had reconfirmed their high regard for the military as an institution and had "a great deal of confidence" in military leaders. (*New York Times,* April 24, 1983, A-24) With such a disparity of views about the public image of the military, lieutenants in search of a vision of themselves might better consider their reputations closer to home, among friends, colleagues, and family.

In the early 1980's, the army's teaching of professionalism was expanded through new 12-hour blocks of service school instruction, much of which was devoted to professional ethics. After reading the words of Huntington and Janowitz, students were asked to consider whether the military life meets the criteria of a profession. The answer seemed to be yes, to the degree that the professionals share a common agreement on standards of competence, commitment, and ethical conduct. The more extreme notions of sacrificing one's family and isolating one's self from civil society did not seem to be a part of the lifestyle of the professionals teaching the courses.

THE NATION'S SERVANT

The true American soldier has always cherished a vision of himself as a *Servant of the Nation,* an identity that runs counter to the idea of professional isolation from the great politic body. In this sense, soldiers feel deeply about patriotism, a patriotism that extols the American nation as man's best hope for guaranteeing the freedom and peace necessary to man's achieving his great potential. Defense of the nation becomes a great calling. With it comes the duty of protecting the people, their value system, and their material well-being. Therefore, soldiers must conduct themselves in a manner that will build strong bonds of mutual trust between the Army and the people it serves.

If soldiers are to expect the citizenry to give over its sons and daughters, its wealth, and its security to the Army's soldiers, these soldiers must earn its respect for competence and fidelity. Conversely, if soldiers are to undertake the obligation of becoming highly skilled in a complex and dangerous enterprise, they must be able to trust that the people will provide them with adequate material and psychological support.

Sustaining strong bonds of mutual trust between soldier and citizen imposes heavy demands for ethical concern and moral conduct on the part of the soldier. Such demands cannot be met by all soldiers. The Army has a certain distribution of the moral "types" familiar to us all—some Christians and some barbarians, some saints and some cannibals. Despite lofty ideology, the Army often displays more liars and cheaters than it can afford. But many statesmen have observed that the distribution of good and bad in the Army is balanced in a way unlike most walks of life—in favor of more men and women of moral stature. Why? Probably because, in a service that so much involves issues of life and death, everyone is more sensitive to the person who might be looking out for only himself in times of danger.

Sir John Winthrop Hackett has often spoken a British view of the moral qualities of the soldier. This was never better expressed than in a lecture at Cambridge University:

> The military life, whether for sailor, soldier, or airman, is a good life. The human qualities it demands include fortitude, integrity, self-restraint, personal loyalty to other persons, and the surrender of the advantage of the individual to the common good . . . This is good company. Anyone can spend his life in it with satisfaction.

Hackett then asked if other walks of life offer comparable satisfaction:

A man can be selfish, cowardly, disloyal, false, fleeting, perjured, and morally corrupt in a wide variety of other ways and still be outstandingly good in pursuits in which other imperatives bear than those upon the fighting man. He can be a superb creative artist, for example, or a scientist in the very top flight, and still be a very bad man. What the bad man cannot be is a good sailor, or soldier, or airman. Military institutions thus form a repository of moral resource that should always be a source of strength within the state. . . . The highest service of the military to the state may well lie in the moral sphere.

—Hackett in Wakin, *War, Morality and the Military Profession*, 123–125.

Soldiers who have searched for a vision of their military selves have eventually come to grips with the moral demands central to their calling, a task rarely accomplished without extensive reading and reflection. In this search they move beyond the occupation of soldiering, seeking to be exemplary human beings, who also happen to be soldiers. In this pursuit of identity, they go back to the courses they took in literature, and history, and philosophy, and they also turn to the religions of their ancestors. They listen to learned men from all walks of life and read their writings.

They seek new experiences in the world surrounding their military haven, learning new languages or sports, undertaking advanced study in new academic disciplines, traveling, or perhaps acquiring new skills in the arts. Recognizing that the best of military men thus occupy themselves, Sir John Hackett quotes Bertrand Russell: "The performance of public duty is not the whole of what makes a good life; there is also the pursuit of private excellence." Hackett adds, "Both are to be found in the military life. It gives much and takes more, enriching freely anyone prepared to give more than he can get." (Wakin, 104)

THE VISION OF COMMAND

Men and women who seek visions of their military selves probably never lose a sense of having many visions that succeed each other in kaleidoscopic patterns. At times, there may be a picture of the battle commander, followed by the possibility of becoming a soldier statesman and strategist. Then, fleetingly, there is a certain accommodation with the professional image, a degree of gentlemanliness, a pride in feeling responsible for the nation's future, a wonderment about the depths of one's morality, and, hopefully, a steady conviction of one's individuality and excellence.

For most inquiring soldiers, however, each of these has a common element—the image of *The Commander*. Peculiar to the martial art is the concept that all decision making, all action, all expression of moral restraint, is centralized in the person of the Commander. This central thread runs through all the visions soldiers might have of their military selves. It becomes a central vision in its own right.

When the younger Patton said to me, "Read in order to command," he was fully aware that not all soldiers will rise to command positions. Some have little interest; others, too little talent. Many prefer a more academic or specialized military line. However, by focusing on command, the military student is encouraged to consider every aspect of military operations and strategy. It is in the mind of the commander that all specialization, personalities, doctrines, and missions must be integrated into some pattern of united effort. The study of command entails the study of all military life.

Therefore, these commentaries on command are designed to illuminate the attributes and ideals of the Commander, in his many roles as tactician, warrior, professional, leader, moral arbiter, and strategist. Much of the commentary is focused on the books that have guided military men throughout the years, and that are being written today by historians, biographers, social scientists, novelists, and philosophers. Within this literature lie the visions that each inquiring soldier can fit into his own particular personality.

Reading About the Military Personality

Of the books cited in this chapter, those with an asterisk are recommended for the variety of visions of the military personality that are portrayed:

* Martin Blumenson, *The Patton Papers*.
  Karl von Clausewitz, *On War*.
* Edward M. Coffman, *The War to End All Wars*.
* Dwight D. Eisenhower, *At Ease*.
* Ulysses S. Grant, *Personal Memoirs*.
  Samuel Huntington, *The Soldier and the State*.
* Morris Janowitz, *The Professional Soldier*.
* _____ , *The New Military*
* Robert Kemble, *The Image of the Army Officer in America*.
  Basil H. Liddell Hart, *The Remaking of Modern Armies*.
* Elizabeth Longford, *Wellington: The Years of The Sword*.
  John P. Lovell, *Neither Athens Nor Sparta: The American Service Academies in Transition*.
* Morris J. MacGregor, Jr., *Integration of the Armed Forces, 1940–1965*.
* George S. Patton, Jr., *War As I Knew It*.
  Helen Rogan, *Mixed Company: Women in the Modern Army*.
  Sam C. Sarksian, *The Professional Soldier in a Changing Society*.
* _____ , *Beyond the Battlefield: The New Military Profession*.
* Michael Shaara, *The Killer Angels*.
  William T. Sherman, *Memoirs*.
  E. D. Swinton, *The Defence of Duffer's Drift*.
* _____ , *Study of War*.
  Malham M. Wakin, *War, Morality, and the Military Profession*.
* Desmond Young, *Rommel*.

CHAPTER TWO

The Challenges of the Commander

The only prize much cared for by the powerful is power. The prize of the general is not a bigger tent, but command.

—Justice Oliver Wendell Holmes, Jr.,
Law and the Court, 1913.

The vision of one's self as a military commander makes sense only for those who yearn to attain and exercise power. Military command requires a concentration of power in one person—power begotten by unusual legal ordination and energized by the will of a person to wield that power. Commanding is a peculiarly military act, rarely undertaken in civilian pursuits where power is customarily more diffused.

To command is to direct with authority. To command a military organization is to think and make judgments, employing specialized knowledge and deciding what those commanded will and will not do. To command in wartime is to assume responsibility for taking and saving human lives. To command in peace and war is to direct how human beings will conduct themselves towards each other. As such, the commander sets moral standards and sees that they are obeyed. To command, therefore, is to think and decide, to feel and moralize, to act and wield power.

Yet, for all these challenges, the art of command is the least understood of all military phenomena. Military schools have taught more about leadership, management, administration, and mapreading than they have about command. The U.S. Army Command and General

Staff College was long a school for staff officers. The American Army has rated commanders on the same efficiency report form as it has rated personnel administrators, chaplains, and computer programmers. Commanders are still picked from paper records, just as bandmasters are picked for promotion and petroleum engineers are picked for higher schooling. Commanders are often rated by superiors on leadership style or their completion of whimfully-selected tasks, rather than on some universally recognized criteria of command performance. Commanders are shifted from assignment to assignment with greater frequency than their foreign equivalents, as if American units have less need for stability in the head positions of their organizations. Commanders are often given tasks to accomplish that are far beyond the ability of their units to complete. There is great scarcity in official literature on the universal requirements, limitations, preparation for, and execution of command.

We have a glorified picture of a man in command—a giant of a man with a title like CG (Commanding General) or CINC (Commander in Chief). Yet, many do not even have "commander" in their titles. A general officer who is senior enough to direct one of the vast staff divisions, such as Army Operations in the Pentagon or J5 of the Joint Chiefs of Staff (JCS), meets the criteria for being a commander. He makes important decisions, sets a moral climate, and is required to direct with authority so that the organization functions in terms of its mission. At this level, however, the staff planning function may in itself perform the thinking that one man might be able to handle alone at the battalion command level.

In the crisis action system of the JCS, for example, a group of field grade officers makes the original assessment of the crisis, develops courses of action, obtains a decision from the joint commander, works out detailed planning, and then issues instructions for execution of the decision. The general in charge of the staff group must see that its members are properly selected and trained, given accurate guidance, and motivated to stay the course in a frustrating experience of shifting international requirements. He holds command authority to see that they perform well as a team.

THREE VISIONS OF BATTALION COMMAND

Most commanders are not generals; they are lieutenant colonels and colonels in charge of battalions and brigades, depots and hospitals, training centers and research laboratories—and ships and air wings. What is expected of them?

When the younger George Patton named for me his five best battalion commanders, he said the one thing they all had in common was *reliability*. Did this mean reliable compliance with his orders? No, this meant reliable pursuit of the unit's overall mission, using great innovation when necessary. He described how Lieutenant Colonel John "Doc" Bahnsen reduced a sticky Vietcong ambush of one of the 11th Cavalry's Troops by grabbing a mess section and a command post vehicle and attacking at the right spot. Patton sought in commanders a toughminded drive to carry out the mission, using as much skilled creativity as possible.

I asked how reliability could be measured in peacetime. Was it a collection of winning statistics in battalion tests, inspections, maneuvers, and re-enlistments? Patton said he had known senior officers who were foolish enough to surround their offices with statistical charts, and then spend their time threatening and cajoling commanders to beat each others' statistics. He preferred to collect just enough data to indicate unusual success or massive failure across the board. He called this "Bruce Clarke's indicator system," in honor of an old mentor. Patton once noticed that a battalion leading the division in re-enlistments performed poorly in command inspections, maintenance, and training. A little probing revealed that the battalion commander had to be relieved for falsifying reports on re-enlistments. "Not reliable," said Patton of him. "The good reliable commanders get the whole mission done, whether or not you keep statistics on them."

I went to a lieutenant executive officer of an airborne company and asked him how he estimated the worth of his commanding colonels. He told me stories of those who "cared more for their careers than for their units." He said that commanders should be judged on whether they worry more about their people than about themselves. I wondered aloud whether a lieutenant can correctly guess what a senior commander really worries about and what motivates these worries. I thought of the long, historical record of highly successful commanders who treated their minions like serfs, with rages of wild temper, insufferable hours, and impossible working conditions. Yet, the lieutenant was asking no more than that his colonels carry out the "leadership principles" taught in American service schools in the 1970's. He had been taught to prize "selfless service" and the soldier morale which is expected to materialize from the commander who expresses great concern for his people.

The lieutenant later added to his criteria, saying that the problem with his current battalion commander was that he wasted his people's time on the wrong things because of unending bad decisions. This I called the dumbness factor. On the other hand, he thought that his brigade commander was smart, but did not carry out his good ideas by riding herd on his staff and subordinates. I counselled that brigade commanders were often left powerless to exert authority because their generals bypassed them or otherwise usurped their power. Overall, I thought the lieutenant was making good progress in developing a philosophy about command. So far he had accounted for selfishness, dumbness, and weakness.

Lieutenant Colonel Wesley K. Clark had commanded a tank battalion for over two years when I last saw him at Fort Carson, Colorado. He thought commanders should be judged on two levels: how they set and maintained high standards of perforrmance in their units, and how they demonstrated moral leadership in achieving these standards. This was a time when the achievement of unit cohesiveness and teamwork were the buzzwords for competence in command. Clark observed that these were easily achieved if one accepted mediocre standards in maintenance, training, and gunnery. Therefore, he would judge commanders against the highest standards they could achieve in readiness tests, inspections, interviews with personnel, or whatever device he or his superiors might create.

I listened to Clark because he had commanded three different companies and was an operations officer at both battalion and brigade level. He survived five AK47 rounds in Vietnam and had been the honor graduate at the Command and General Staff College. He had taught philosophy to cadets following his work at Oxford as a Rhodes Scholar. Throughout, he had learned to pay very close attention to detail, particularly to the details of military equipment and administration—a habit found in nearly all great military commanders from Napoleon to Rommel. Clark felt that knowing the details allowed creativity to replace "cooky-cutter" solutions to problems, solutions that were never quite right for the situation at hand. He also felt that attention to detail allowed a commander to sense oncoming problems while there was still a chance to head them off. Overall, Clark measured battalion commanders on their ability to achieve high standards of wartime readiness, while sustaining moral standards in their daily acts and statements.

THE SEARCH FOR CONSENSUS ABOUT COMMANDERS

Patton, the lieutenant, and Clark were all correct about how battalion commanders should be judged. But each saw the problem from a different perspective—Patton from above, the lieutenant from below, and Clark in the job at hand. Does this mean that there are no absolute criteria available to the disinterested, impartial observer? Or to the officer making out efficiency reports on his commanders?

The designers of efficiency report forms attempt to isolate meaningful criteria for the rating of all Army officers, and thereby of all commanding officers. In the late seventies, Colonel Dandridge "Mike" Malone studied and summarized a dozen efficiency report forms used in the armed services of the United States, Canada, France, and Britain. He was in search of a consensus on the traits expected of the military officer, including commanders. There was greatest agreement (on 10 of 12 forms) on the need for *judgment* and *initiative*, qualities associated more with the mental process of thinking and deciding, than with physical or spiritual aspects of military life. The consensus was also extremely high on the need for *professional knowledge, good management of resources,* and *oral and writing ability.*

Having dealt with such cerebral processes, the designers of these efficiency reports placed priority on *cooperation, force, moral courage, responsibility,* and *looking after the welfare of subordinates.* Of some 47 traits reflected in these reports, selflessness was mentioned only in recent U.S. Army reports, and *loyalty* prized only in American Army and Navy evaluations. Overall, there was close consensus on the importance of judging commanders on how they think. For me, however, there appeared to be a certain faddishness for "buzzwords-of-the-day" in the remainder of the traits hoped for in military commanders.

Historians, biographers, poets, and novelists seem to have a better grasp than designers of efficiency reports of what makes a commander good or bad. On the first page of *The Masters of the Art of Command,* Blumenson and Stokesbury describe the traditional image of the great commander as "knowledgeable in his profession, experienced, bold, brave physically and morally, an impressive man of decision and action." In touching on this trilogy of mental, moral, and action-oriented concerns, they may well have drawn upon J.F.C. Fuller, the twentieth century's most diligent biographer of great commanders.

In his monographs on the generalship of Alexander, Caesar, and U.S. Grant, Fuller outlined the mental, moral, and physical traits necessary for commanders. From the brain of the great commander, Fuller expected imagination, independence of thought, and accurate insights into what the enemy commander would do. From the heart he expected physical and moral courage and, with it, heroism in battle. The commander's courage, he wrote, provided the "moral cement" that we today call "unit cohesiveness." Finally, as part of the "personality factor," Fuller expected great energy from the commander, who was ever present in all units on the battlefield, even if it meant 50 miles every day on horseback.

In his writing on generalship and command, Fuller drew upon and quoted extensively from Napoleon, Marshal Saxe, and Clausewitz. He suggested a certain historical continuity in the nature of command, regardless of changes in weaponry, communications, and styles of wielding power. This ageless vision of the command function can be sensed among nearly all the modern military biographers, starting perhaps with Colonel G.F.R. Henderson's *Stonewall Jackson and the American Civil War* (1895), and seen most expressively in Douglas Southall Freeman's *R.E. Lee* and *Lee's Lieutenants*. The twentieth century military biographers have been particularly insistent on applying a moral measure to the commander, in terms of how he treats his people, noncombatants, and his enemies.

It goes without saying that this consensus among historians and biographers vanishes when they begin to argue over what constitutes a good decision, what is and is not moral, and whether the qualities of impetuosity, stubbornness, aggressiveness, and other symptoms of dynamic character are virtues or vices in a given action. After years of researching commanders from the Middle Ages to World War II, Barbara Tuchman concluded that their primary quality was *resolution*. That is, "the determination to win through, whether in the worst of circumstances merely to survive, or in a limited situation to complete the mission; but whatever the situation, to prevail." She praised General Joseph Stilwell's command presence in World War II Asia as having "the absolute, unbreakable, unbendable determination to fulfill the mission no matter what the obstacles, the antagonists or the frustrations." (Tuchman, *Practicing History,* 278) This same resolution is often disastrous, of course, in a general hell-bent on the wrong objective while decimating squadrons of good troops, as seen in the British commanders who ordered the charge of the light brigade.

Biographers have difficulty with the "I would rather be lucky than right" syndrome. What is one to say of the proven commander whose attack gets snowed on, allowing his dull-witted, depraved opponent to come in with a smashing victory? Even the idea of victory itself gives constant problems; some great commanders get assigned to hopeless theaters of war, unable to win enough to inspire their troops to greatness. Only a Napoleon, when confronted with such a fate in Egypt, could abandon his army and order himself back to Paris to contrive more fruitful battlefields.

The military novelists have a better time, for they can invent the circumstances under which their hero or buffoon reveals his talents and ethics. The novelist can also cause his commander to elaborate on his thoughts and emotions, while the historian veers from the documentary record only at great peril. When C.S. Forester, however, carves out the personality of Horatio Hornblower in *Beat to Quarters,* or Herman Wouk that of Captain Queeg in *The Caine Mutiny,* or Anton Myrer of Sam Damon in *Once an Eagle,* none of them move far beyond the historian's consensus about the mind, moral fiber, and the will-to-action of the model commander. In fact, by inventing the situation through which their commanders display their skills (or lack of them), the novelist can offer a clearer and more impressive lesson to the reader than can the historian enmeshed in the complicated historical truth. For those who wish to command, the military novel is a great place to start learning.

THE ELEMENTS OF COMMAND

The art of commanding military units is one of the few human endeavors that cannot be learned by taking courses leading to a degree in "Commandership" or "Commanderism." This good fortune can be attributed to the fact that command is very personal. It is so much an intermeshing of personality and circumstance that the best teachers abjure the thought of finding generalizations or rules that can be applied to all commanders in all situations. They do not rule out the memorization of guidelines such as the Estimate of the Situation, for analyzing military problems, or the Five Paragraph Field Order, for issuing instructions to subordinates. But, they also show how commanders use them differently, describing many kinds of commanders in many kinds of circumstances, and warning their pupils that they must find their own styles and techniques.

It is fortunate, therefore, that the most demanding command experience does not confront the soldier until he is well over thirty and has had at least a decade of military service. Lieutenant platoon leaders are just that; they *lead* their troops in accomplishing missions that are created elsewhere and passed down to them for execution. They have no staff; the moral climate is set for them more than they set it.

A captain commanding a company held a true command in the Old Army, and does so now in wartime. In today's peacetime Army, however, the power of the company commander has been denigrated by modern communications, by theories of management that have moved much of the company's administration to higher headquarters, and by centralized systems of pay, promotion, training, maintenance, and supply that bypass the company commander's authority and impact directly on the soldier below him. It is possible for a captain of average ability to be quite successful in the eyes of higher authorities if he faithfully obeys orders, enforces standards set by others, and does not violate some cardinal rules of leadership and management. This is good followership, but it is not command.

At the level of battalion command, however, there is no evading the responsibilities of command. The very scope of planning and directing the activities of some 600 individuals requires the use of a staff and a battery of subordinate commanders. The lieutenant colonel commanding a battalion must create priorities, identify objectives, allocate resources, dictate timing, interpret missions, and undertake all the tasks associated with military command, both in peace and war. His responsibilities are much more akin to those of a brigadier general than to those of a lieutenant or captain.

At the highest levels of command, the world changes once again. In *On War,* Clausewitz wrote: "There are Field Marshals who would not have shone at the head of a cavalry regiment, and vice versa." Today, it is just as dangerous to assume that those who make good battalion, brigade, and even division commanders will make good senior commanders at the three and four star rank. This is a third and quite distinctive level of command, where the familiar world of the infantryman or the aviator seems to disappear into the past. The very senior general must direct activities of units and forces about which he has little familiarity and, sometimes, exceedingly little knowledge. He begins calling upon all his creative insights and moral estimations and conceptualizing skills in order to cope with his new agenda. In *The Professional Soldier,* Janowitz described the very senior generals as a small elite within an elite, made so by their independent-minded ability to rise above conventionality and established doctrine. B.H. Liddell Hart said of senior generals:

> Creative intelligence is and always has been the supreme requirement in the commander . . . coupled with moral character. . . . The best hope of tilting the scales and of overcoming the resistances inherent in conflict lies in originality—to provide something unexpected that will paralyze the opponent's freedom of action.

—Liddell Hart, *Thoughts on War,* 123.

Whether today's senior commanders can be innovative in an era of swift change in military technology and weaponry will depend on the quality of information they receive and the responsiveness of the systems they command. Martin van Creveld, for example, suggests in his 1985 *Command in War* that 4-star generals who are winners will probably push operational decision-making down to the 2-star division commanders, who are better positioned to make such decisions.

To command at any level is to do more than just manage military forces, if management means "working with and through individuals and groups to accomplish organizational goals." (Hersey and Blanchard, *Management of Organizational Behavior,* 3) The vast literature of management theory can help the commander in the process of planning, organizing, controlling, and coordinating his unit, especially in a peacetime environment. Just as a commander changes his methods as he moves from junior to senior to very senior levels of command, so the manager adopts new ways on the path from supervisory to middle to top management levels.

Management theory has its limitations, however, because the research that supports it is based almost wholly on civilian institutions. Management literature is virtually silent on the commander's requirement to shoulder 24-hour responsibility for "employees" whose livelihood and motivation depend substantially on federal law and bureaucracy. The commander will not find in management theory the insights and values that can explain to soldiers why their organization is more important than they are, why it can be sacrificed to national need, and whether they may live or die in the process. These differences between command and management, as well as those between management and leadership, were recently highlighted in *Military Leadership: In Pursuit of Excellence,* edited by Taylor and Rosenbach of the U. S. Air Force Academy.

To command is to do more than lead soldiers, if military leadership is "a process by which a soldier influences others to accomplish the mission"—the most recent Army definition. The eyeball-to-eyeball contact necessary for leadership at squad, platoon, and company levels begins to wane at the level of battalion command, and virtually disappears at higher levels. Senior commanders spend most of their waking hours deciding what the mission will be and setting guidelines for execution by subordinates. When spending time with those who are executing the mission, their purpose is to adjust the mission and prepare for the next decision. If "influencing others" is necessary, it is usually in the direction of applying pressure to the faint hearted and the confused. This, too, has its limitations; many Vietnam veterans talk with bitterness of commanders hovering over the battlefield, broadcasting appeals and threats, while interfering unduly with the ability of junior leaders to get the job done. Among senior commanders, the yearning to "influence others" must be carefully rationed.

To command is to do more than carry out orders and apply rules and regulations to the ebb and flow of military administration. Command calls for a creative act, spawned by a carefully carved vision of one's mission and professional values. Great commanders have the confidence and courage to interpret rules and orders, and to put their personal stamps on the decisions guiding their forces. U.S. Grant said almost contemptuously of his peers:

> if men make war in slavish observance of rules, they will fail. . . . While our generals were working out problems of an ideal character, problems that will have looked well on a blackboard, practical facts were neglected. To that extent I consider remembrances of old campaigns a disadvantage. Even Napoleon showed that, for my impression is that his first success came because he made war in his own way, and not in imitation of others.

—J.F.C. Fuller, *Grant and Lee,* 82.

To command, then, is to manage well when management is called for, to lead well when leadership is necessary, and to carry out orders and enforce regulations when "going by the book" is all that is required. But to confuse each of these three activities with the full scope of the command function itself is to underestimate the need for taking intellectual and moral responsibility in the performance of one's military duties.

The Vietnam War taught the American Officer Corps how vast a vision a commander needs in a war with no front lines, an unidentifiable enemy, a mix of political and military objectives, a hostile journalistic fraternity, and questionable support from the American people. One could well start a study of command responsibility with General William C. Westmoreland's *A Soldier Reports,* a view from the Man in Charge and Douglas Kinnards's *The War Managers,* an analysis of how 60% of the generals who served in Vietnam viewed the successes and failures of what they did.

In the aftermath if Vietnam, however, we find the same misunderstandings about command that existed before. There are general officers who expect their colonel commanders to be little more than hyperactive platoon leaders, standing before their troops and exhorting them on to greater glories, as at a pep rally. Others expect their commanders to be comfortable cronies, fitting their activities to the whims of the boss, regardless of the real needs of the organization or troops. For these generals, command failure is equal to lack of "leadership" or lack of loyalty.

On the other hand, when competent generals ask "How do I know when a colonel is failing, or has failed, his command assignment?" they usually find the answer in one of the following conditions:

- When his decision making squanders the available time, manpower, and resources.

- When he is not accomplishing his many missions, even though he has adequate resources.

- When he is not training his forces to a high level of technical and tactical competence.

- When he is not establishing a moral climate that provides for responsible behavior of his troops, and for justice for both his people and the Army.

- When he is not in control of his people and his organization, as evidenced by their failure to carry out his own orders.

In short, the astute generals look for signs of good thinking and decision making, a sensitivity to moral and ethical requirements, and the resolute actions necessary to insure command success.

LEARNING ABOUT COMMAND

From the day a soldier is commissioned, he begins to learn how to handle his first real command. Once there, he intensifies his study until he will no longer command. Three kinds of learning go on simultaneously—specialization, professionalization, and human growth. Military specialization is acquired in service school Basic Courses, followed by several years' emersion in one's commissioning branch; additional specialties may be added later, including understanding of naval, air, and allied forces. Learning about the profession, which binds together all the specialties, begins with orientations in early years and develops through combined arms and staff training at the Command and General Staff College.

The learning that expands one's human potential prepares soldiers for the senior levels of command, where responsibilities often encompass organizations and missions left unexplored in specialist and professional training. This is the learning that is thought necessary for all responsible human beings; it transcends purely military affairs and allows the commander to see his troops as human beings and his work as an expression of all human endeavor. This learning starts in one's general education in secondary schools and is developed as an undergraduate and in personal reading and study. Continuing development of this human potential can arm the senior commander with a well integrated personal philosophy, a reservoir of moral conviction, and a vision of what is achievable as well as what is desirable. The essentiality of these traits among senior commanders explains why the gifted amateur often succeeds—like Julius Caesar and Oliver Cromwell—and, even more often, the dedicated professional fails.

Military historian I.B. Holley of Duke University says that skilled learners pursue four objectives in any learning exercise they undertake; knowledge, skills, insights, and values, as defined in Chart 2. Knowledge and skills are at the heart of specialty and professional training. Military schools generally impart knowledge of weapons, equipment,

Chart 2

Four Learning Objectives for Commanders

---

1. *Knowledge.* Information, data, facts, theories, concepts. Answers question: "What should I know?" May be achieved by many learning methods. Highly perishable.

2. *Skills.* Abilities that can be developed and manifested in performance, not merely in potential. Answers question: "What should I be able to do?" Includes technical, communications, information-retrieval, and some analytical skills.

3. *Insights.* Ideas and thoughts derived internally from an ability to see and understand clearly the nature of things. Necessary part of making judgments, of deciding, of "putting it all together," of "being aware" of wisdom, far-sightedness. Answers questions: "What does this mean? What is important in this situation?" Cannot be taught directly, but can be induced by qualified teachers. Generally a product of education rather than training.

4. *Values.* Convictions, fundamental beliefs, standards governing the behavior of people. Includes attitudes towards professional standards such as duty, integrity, loyalty, patriotism, public service, and phrases such as "take care of your people" and "accomplish your missions." Answers questions: "What do I believe? Where do I draw the line?" Values, like insights, must be derived by the individual, if they are to have meaning.

---

procedures, and ideas useful to commanders, such as the double envelopment or the principles of war. Training schools can also engender technical and communication skills and, if they have strong educational credentials, can give the future commander a leg up on those skills that conceptualize and analyze problems.

But it is a different matter for a Patton or Rommel to acquire those precious *insights* that tell him what is important in a given situation, or for a Marshall or MacArthur to come to prize those *values* that made men follow him in pursuit of great national victories. Such insights and values must be self-induced, whether from experience or from reading and education. Or they may be engendered with the help of a great teacher, such as Eisenhower had in Fox Conner, Napoleon in de Guibert, and Philip of Macedon in Socrates.

In the ongoing renaissance of the U.S. Army Command and General Staff College, which took full effect in the 1980's, several important steps provided institutional support for commander learning of essential insights and values. The Center for Army Leadership was founded, to provide more structured learning in professional ethics and related leadership competencies at all Army schools. The School of Advanced Military Studies was established, to provide selected officers with a full year of exploration into military history and biography, linked to modern problems of the operational art and the AirLand battlefield.

Recognition of the special problems of the commander was dramatized most clearly, however, in the institutionalizing of Pre-Command courses for all Army selectees for battalion, brigade, and division command assignments. Within the new School for Professional Development at For Leavenworth, these courses were fashioned in the mid-eighties by Col. Robert F. Broyles and Lt. Col. Richard C. Stubbs, with an emphasis on diagnostic assessment of command selectees, followed by instruction in areas of greatest need to them. Orientations by senior generals on current policies were coupled with classes on tactics, administration, military law, professional ethics, and the status of current technological advancements. Individual attention was given to each commander's upcoming assignment. Commander spouses attended Command Team Seminars, exploring the needs of Army families across the world. Correspondence back to the Pre-Command Course from the field indicated that the students greatly appreciated the instruction and confirmed that constant updating was necessary to keep the School abreast of continuing changes in the Army.

The introduction of these new learning enterprises broke the pattern of Leavenworth being a college designed more for staff officers than for commanders. Attention given to the command function was causing a subtle realignment in the three levels of officer learning; the traditional hierarchy of junior, field grade, and general officers was bowing to a new hierarchy—junior (lieutenants and captains), senior (lieutenant colonels through major generals), and very senior (three and four star generals). Differentiations of this type more clearly matched the points where officers must shift gears mentally as they move up through three levels of command, management, and leadership. Finally, the renaissance at Leavenworth pointed the way to increased self study and extension study outside the schoolhouse, in recognition that learning must be continuous in a continuously changing world.

STUDYING THE COMMANDER'S ART

When George Patton and Wesley Clark and the lieutenant talked to me about command, each was aware of the many roles of the commander—tactician, strategist, warrior, ethicist, leader, manager, and technician. Each sought ways to mesh his own personality with the many facets of the command process. Each had his own conception of the relationship among command, leadership, and management. Each was a voracious reader, and maintained a sizeable library.

Reading about the command function is normally begun with study of a standard work in the history of military art and science, of which Maurice Matloff's *American Military History* is the most well-known, and the *West Point Military History Series* the most complete. Thereafter, the inquiring soldier fits his reading to his own interests, mixing monographs about command with biography, memoirs, plays, and novels. A list of some outstanding books in these categories appears at the end of this chapter.

The best analysts of military command have had military experience, but may have been distinguished more by their writing than by their reputations as commanders. An easily identifiable exception is Field Marshall Sir Archibald P. Wavell, whose 1939 Cambridge lectures, "Generals and Generalship," were of such interest to Rommel that he carried them with him throughout World War II. After the war, Frau Rommel found the worn book in his field gear, and sent it to Wavell's wife in England. These lectures were subsequently reprinted in the 1953 collection of Wavell's writings, *Soldiers and Soldiering,* which is cited below among the best of military monographs and lectures.

Although good military biographies lie at the heart of any study of command, it is difficult to find the best among so many third-rate military books. The biographies cited below go beyond a mere retelling and description of events, and provoke the mind with questions and analysis. J.F.C. Fuller explains his philosophy of the commander's art in writing about Alexander, Caesar, and Ulysses S. Grant. Liddell Hart argues back in studies of William Tecumseh Sherman and other Great Captains.

One is often forced to choose between the long work, such as Forrest Pogue's three-volume *George C. Marshall,* and the short book, such as Leonard Mosely's *Marshall: Hero For Our Times.* Often, the shorter overview provides a basis for deeper selective study in the definitive work. In the one-volume condensation of the Freeman *R. E. Lee,* however, one loses the footnotes and commentary that provide the best insights. Professional historians praise D. Clayton James' multi-volume *The Years of MacArthur.* They give mixed acceptance to the William

Manchester portrayal of MacArthur in *American Caesar* as unduly biased.

Memoirs serve many purposes for both their authors and their readers. Sir William Slim's *Unofficial History* entertains as it cleverly instructs subalterns and brigadiers on how to survive Middle East bullets and politics. William Tecumseh Sherman's *Memoirs* are again becoming fashionable reading in the study of professional ethics.

In addressing the problems of the commander, the novelist has the opportunity to design a situation not too far from reality, which can bring about the best and worst in his characters. The story often illuminates the commander's dilemmas better than real life does, with its muddy eddies and murky truths. A classic example is *The General,* a composite of British commanders in World War I, as seen through the eyes of C. S. Forester.

In reading about the lives and thoughts of real or fictitious commanders, serious students find certain questions worth pondering or discussing in seminar: What are the essential elements of this commander's knowledge, skills, and values? Do these change as he moves from junior officer to senior general? How did he acquire these assets? How creative has he been? How does he translate his ideas into action? What is the basis of his authority and how does he use it?

Is this commander adept at all command roles—leader, manager, tactician, trainer, strategist, warrior, and moral arbiter? What has been his vision of himself as a military person, and has he achieved that vision? Is he an appropriate model for me, or are his style and personality too different from mine?

The soldier who would command reads for the insights and values of those commanders who went before him—for the manner in which they made decisions about what was important and how it could be obtained. Above all, one reads for those qualities of character that General George C. Marshall underscored in his talk to the first graduating class of the Officer Candidate School at Fort Benning, Georgia, in September of 1941:

> When you are commanding, leading men under conditions where physical exhaustion and privations must be ignored, where the lives of men may be sacrificed, then, the efficiency of your leadership will depend only to a minor degree on your tactical ability.

It will primarily be determined by your character, your reputation, not much for courage—which will be accepted as a matter of course—but by the previous reputation you have established for fairness, for that high-minded patriotic purpose, that quality of unswerving determination to carry through any military task assigned you.

Reading About Command

1. Useful monographs, lectures, and other analyses of the command function, some of which were cited in this chapter:

   Karl von Clausewitz, *On War*.
   Martin van Creveld, *Command in War*.
   J.F.C. Fuller, *Generalship: Its Diseases and Cures*.
   Thomas E. Griess, ed., *West Point Military History Series*.
   Paul Hersey and Kenneth H. Blanchard, *Management of Organizational Behavior*.
   Morris Janowitz, *The Professional Soldier*.
   Douglas Kinnard, *The War Managers*.
   Basil H. Liddell Hart, *Thoughts on War*.
   Maurice Matloff, *American Military History*.
   Robert L. Taylor and William E. Rosenbach, eds., *Military Leadership*.
   Barbara Tuchman, *Practicing History*, "Generalship."
   Field Marshall Earl Archibald P. Wavell, *Soldiers and Soldiering*.

2. Biographies and memoirs with analysis of the command function:

   Alexander: J.F.C. Fuller, *The Generalship of Alexander the Great;* Mary Renault, *The Nature of Alexander*.
   Caesar: *The Conquest of Gaul;* J.F.C. Fuller, *Julius Caesar: Man, Soldier, Tyrant;* Michael Grant, *Julius Caesar*.
   Napoleon: Albert S. Britt, *The Wars of Napoleon;* David Chandler, *The Campaigns of Napoleon;* James M. Thompson, *Napoleon Bonaparte*.
   Grant: *Personal Memoirs;* J.F.C. Fuller, *The Generalship of Ulysses S. Grant;* William S. McFeely, *Grant: A Biography*.
   Lee: Douglas S. Freeman, *R.E. Lee* and *Lee's Lieutenants*.
   Jackson: G.F.R. Henderson, *Stonewall Jackson and the American Civil War*.

Sherman: *The Memoirs of William T. Sherman;* Lloyd Lewis, *The Fighting Prophet.*

Bradley: Omar N. Bradley, *A Soldier's Story.*

Marshall: Forrest C. Pogue, *George C. Marshall,* 3 vols.; Leonard Mosely, *Marshall: Hero for Our Times.*

Patton: George S. Patton, *War As I Knew It;* Harry H. Semmes, *Portrait of Patton.*

Stilwell: Barbara Tuchman, *Stilwell and the American Experience in China.*

Slim: Field Marshall William Slim, *Unofficial History.*

Montgomery: Nigel Hamilton, *Monty,* 2 vols.

Ridgway: *Soldier: The Memoirs of Matthew B. Ridgway.*

Westmoreland: *A Soldier Reports.*

Collective biography: Martin Blumenson and James Stokesbury, *Masters of the Art of Command;* J.F.C. Fuller, *Grant and Lee;* Edgar F. Puryear, Jr., *Nineteen Stars;* T. Harry Williams, *Lincoln and His Generals.*

3. An annotated listing of novels and plays that delineate the reaches of the command personality:

James G. Cozzens, *Guard of Honor.* "The best novel of World War II" portrays command in a small Army Air Corps base.

C.S. Forester, *Beat to Quarters.* The first of ten books on fictitious Captain Horatio Hornblower's command of British ships in the Napoleonic era.

_____ , *The General.* Hitler had his generals read this portrayal of an heroic incompetent in World War I.

William W. Haines, *Command Decision.* A play and movie about decisions to spend lives in World War II.

Paul Horgan, *A Distant Trumpet.* Command responsibility in the war with the Apaches.

Anton Myrer, *Once An Eagle.* How Sam Damon commanded, from War I to Vietnam.

Michael Shaara, *The Killer Angels.* How commanders probably formed their decisions at the battle of Gettysburg.

William Shakespeare, *The Life of Henry V.* "Once more unto the breach, dear friends, once more."

Herman Wouk, *The Caine Mutiny.* Only the lawyer understood the effects of command responsibility on Captain Queeg.

CHAPTER THREE

The Company Commander

I hope you will graduate with an appreciation of the transition that is necessary to go from leadership to commandership. . . . It's an important thing and a lot of people have never bridged it. They're still exercising leadership as company commanders or even when they get higher, and by doing so they're bypassing or poorly using their subordinate commanders and staffs.

—General Bruce C. Clarke
to USACGSC Class of 1963.

After his retirement as Army Chief of Staff, Matthew B. Ridgway wrote that the power of the company commander had declined during his long career, especially in the decade after World War II. He cited the loss of the commander's right to promote sergeants as one of the many "reforms" that had undermined the company commander's authority. (Ridgway, *Soldier*, 21.) This trend accelerated into the seventies. Some of it was dictated by new realities, such as a greater need to closely control more lethal weapons. But most of this usurpation of power came from the conscious decisions of senior commanders. They wanted more direct control of the lower echelons of an Army that was scattered across oceans and continents, subject to continuous political and media scrutiny, and too frequently led by officers of marginal talent and training.

By the late seventies, the company commander was involved in such directives as where, when, and how to train, maintain, supply, pay, feed, and lead his force according to the general's wishes. His job was to *lead* the troops in prescribed directions, with standards and conditions established by superiors rather than him. The power to *command* was retained at battalion and above, for it was in those realms that creative decision making was located, along with the power to establish the moral climate that would prevail.

The effects became apparent in the Vietnam War, when control began to slip away in the units most distressed by excessive personnel turnover, racial disharmony, inadequate training, low standards of discipline, and marginal quality of recruiting. Books such as *Crisis In Command,* written in 1978 by Gabriel and Savage, described a breakdown in the necessary cohesiveness of some combat units, blaming Army policies based on civilian management philosophy. While many disagreed with this analysis, the arguments had enough merit to cause in the 1980's, a revival of such antique concepts as the "regimental system" and unit replacement of personnel.

Interviews with captains who had commanded companies in post-Vietnam, Germany, Korea, and the United States indicated that a wide variety of styles and techniques were being employed. Some said they had to adjust to a rigidly regimented environment, in which they had little discretion in handling personnel, training, and readines requirements. Others, however, said that they had all the power and independence of action that they could handle. The difference seemed to spring from the command philosophy of their superiors, and from the type of mission they were expected to carry out.

THE ANDERSEN COMMAND

Captain William E. Andersen commanded two companies in very different environments. In the first, he felt he had sufficient power and authority and was truly in command; in the second, he felt he was little more than part of a system largely outside his control. The first was in a stateside airborne division, where his superiors wrote of him: "The finest company commander I have ever known. . . . could be a field-grade officer today. . . . he relates to all ranks. . . . [and] has the complete respect of every soldier in his company and the battalion." While he was in command for a year, his company was praised for developing the ground tactical plan for air assault, live fire exercises in airmobile units; was cited for receiving the highest ratings in the Jungle Operations Training Course in Panama, where they practiced two air deployments; and was commended for exceptional performance in unannounced maintenance inspections and for scoring 100% on the Annual General Inspection. Most specifically to the point of my inquiry, he was credited for "a rise in company cohesion, esprit, and self-confidence." (Quotations from Andersen's Efficiency Reports.)

I asked Andersen to explain how he went about this command assignment. He told me, "I was still a Lieutenant in February, 1979, when I took command of B Company, 1/503rd Infantry, of the 101st Airborne Division (Air Assault) at Fort Campbell, Kentucky. My commission was only two and a half years old and I was to be the youngest infantry commander in the division. That is why it is so surprising that I was given so much authority. It came from a very unique chain of command—my battalion commander Lt. Col. David A. Bramlett, one of the very best, and the 3rd Brigade Commander was Col. Peter M. Dawkins. They more often than not worked for Brig. Gen. William B. Louisell, who had a reputation for experimenting with new ideas.

"I followed a few general rules for commanding the Bravo Bulldogs. First, you (the Commander) have to be more technically competent than anybody else on weapons, tactics, communications—all that technical knowledge that makes you the company's most expert trainer. Second, you must seize the initiative with higher headquarters every day, become the planner of your outfit, and help them avoid sending you off in wasteful directions. Third, you have to eliminate some poor quality soldiers, so that the 10% bad element does not prevent the other 90% from doing a good job.

"The bad news was this: My company was authorized 157 soldiers, but rarely had more than 115 at any given time. Eventually we operated three rather than four platoons. But we still had the full complement of weapons, so it took more than a day to clean weapons, when normally it should take no more than a few hours. We also had an eighty percent turnover of soldiers during the year, which meant that a lot more time went into administrative work, at the expense of training. We put a lot of quarters into the machine, which provided no pay-off in unit effectiveness.

"My people came mostly from backgrounds of economic and social poverty. It is beside the point that 60% were black or hispanic. Nor is it very important that 50% did not complete high school. We have excellent soldiers—some of our best sergeants—who are from minorities and are undereducated. Rather, it is important that half had major personal problems—indebtedness, court records, jailings, kicked out of their homes, destitute parents. The majority had been losers all their lives; never real winners with girls, sports, school, cars, or anything important to young people. Most important, many were without ambition, hope, or any sense of personal discipline.

"We had three kinds of soldiers coming and going through the company. About 60% were basically good soldiers, and 10% were unredeemably poor soldiers. In between was that 30% borderline group who could go either way. I had to give the good soldiers conditions under which they could thrive best, and count on their winning over the borderliners. This meant that I had to do something about the bottom 10%.

"Some of the worst soldiers should never have been enlisted, having no ability to adjust to a new life because of mental or aptitudinal ineptness; they could be given honorable discharges. There were also those who were 'unsuitable'; repeated troublemakers who could be given a general or 'other than honorable' discharge. Men who were less damaging to the Army could be barred from re-enlistment at the end of their current term of service. When I flagged their file as 'barred,' they were blocked from promotion, bonuses, rewards, or citations while still on duty, and this action followed them in any transfer to another unit. Barring and discharging were powerful incentives for men of dubious soldierly qualities to follow my lead. During my year in command, I ushered about 20 bad soldiers out of the Army (or about 10% of those who came to Bravo Company). As for barring, a snapshot on any given day would show that about 15% of the company's soldiers were on administrative hold until their three-year enlistments were completed.

"Why pursue this so aggressively when the Army is short of manpower? I did it *because* the Army is short of manpower and you have to get the most out of what you have. If a few soldiers are allowed to remain in the unit while they refuse to clean weapons, smoke pot and get stoned while on guard, and push dope to new soldiers, the borderliners will join them and the good soldiers will become cynical and quit working. When you spend your weekends preparing cases for "expeditious discharge," you back your good people and deter others. You also exert the command authority that is expected of you—they know who is boss, and that pays off in the field and other times when the going gets rough.

"I was once told that my most important job was to build a company that would win in combat. You can't do that with people who feel that they are losers. It is important that they win enough to begin to think like winners. How? Here's an example: We had a division policy that gave a company a day off for every month there were no soldiers reported absent without leave (AWOL). But a few soldiers create the AWOL statistics, which prevent other soldiers from winning that day off. After I eliminated about six leading 'AWOLers', we went eight months winning days off. The troops became very proud of that and kept each other from going astray.

"In the same way, after we discharged some of those who quit on everything, it became possible to win the War Eagle Sprint, and to become the best running company in the brigade. For six months we ran at the head of the column with our streamers, bulldog shirts, and B.S. Bulldog, the mascot. Any company is a potential winning company, but it can't be done if you let a handful of bad soldiers drag the others down."

A COMPANY COMMANDER'S FOUR MAJOR CONCERNS

Whether or not Captain Andersen was right in his winners-losers theory, he was commanding in the tradition of the best military leaders. Units are organized so that the efforts of all individuals together are greater than the sum of what each individual could accomplish alone. If this is to happen, there must be a cohesiveness of thought and action, in which the importance of the unit exceeds that of the individual. If soldiers are to give the unit that necessary part of themselves that will provide such cohesion, they must feel that they will gain, in return, something very important to them, such as comradeship, trust, security, and a certain pride of accomplishment. By giving the soldier reason to believe that not everyone was good enough to be in Bravo Company, and by winning rewards for them, Andersen was building that essential unit cohesiveness that makes military organizations succeed.

Discharging and barring was only one part of Andersen's program for giving good soldiers reason to devote themselves to their company. He also used an elaborate system of certificates, ribbons, and medals. He had one of the highest promotion rates in the division. There were monthly unit parties; the barracks were completely repainted after reconstruction of dayrooms and gamerooms. Individual rooms got new furniture and drapes. Perhaps most important, there was a restructuring of activities so that soldiers could begin to see some order in the chaos that can become daily Army life.

As we talked about Andersen's rules for creating a cohesive company, it became clear that he focused on four major concerns: correctly designating unit goals, demanding high standards in discipline and performance, developing subordinates, and, as he put it, "making certain that the soldiers understand that you are interested in their well-being."

As for designating goals, he warned that the little time and resources available must not be fragmented, but aimed carefully toward a few priorities. "You must decide what you want, then fit it in with the desires of your superiors. This means a lot of time explaining to them and their staffs, in their offices and on the telephone. Then you must carefully explain to the soldiers what they have to do in the immediate future, in briefly stated, achievable objectives that become rallying points for the unit. In units I had served in as a platoon leader, the commanders rarely told the troops clearly what the goals were. In mine, I would like to frequently give every soldier—always every squad leader—a piece of paper that gives him his mission for the next round of activities. The copying machine is your best friend."

What standards of performance should we set for soldiers who clearly understand the unit goals? Andersen thought that people made a mistake in assuming that commanding less-able recruits meant setting standards low enough for them to achieve. "In order to have winners, there has to be something worth winning, and you have to work hard for it. I was not popular at all when I began demanding performances of duty and conduct that were more exacting than in neighboring units, or in the units my soldiers had served in before. But then they began to win, and I began to reward them with promotions and medals. After they worked the company into a maximum score on the Annual General Inspection (AGI), I handed out 4 immediate impact Army Commendation Medals and over 50 of my Bulldog achievement certificates. They prized these symbols of winning—put them on their walls, sent them home.

"The other side of the coin is the punishment policy for failure to perform—discharging and barring for the worst, and an occasional court martial. Early in my command I gave Article 15 punishment about 5 times a week—reduction by one grade, loss of 5 days pay, 14 days extra duty, a week in the post correctional facility, things like that. Once we settled down, I needed to use Article 15 no more than once a week. It did not take long to establish that I accepted only winning standards.''

I pressed Andersen on the indispensability of having excellent non-commissioned officers if one is to have a cohesive combat unit. He agreed that he was very dependent on an extremely competent First Sergeant, James R. McCosh. ''But, you must pass responsibility down to platoon sergeants and squad leaders, even if they have a long way to go in developing technical competence and that necessary concern for the welfare of others. Only when they are held responsible for missions will they begin to feel responsible for the well-being of the unit. Again, you must tell them in writing what their responsibilities are. I sometimes used blackboards and those big wall tablets in the hallways.

''You must also reward them for doing right—they have to learn how to become winners, too. I had one superb E2 who had a good civilian background in business. I promoted him through E3 and E4 to Acting E5 in five months, as a model for others. These young noncommissioned officers need a lot of backing. In the field, senior officers too often berate and humiliate them for obvious mistakes. You must be there to stand between that ranking officer and the sergeant who is trying to learn. Overall, I can't think of any surer way to unit cohesiveness than concentrating on developing leaders out of people who are groping for some way to become better men.''

Andersen said that he placed great emphasis on soldiers' understanding his concern for their well-being. ''If a soldier is to volunteer an important part of himself to advance the good of the unit, he must feel that his commander is technically competent and that he is telling the troops the truth. Technical competence—it shows when you spend a lot of time getting the company the right manuals and showing the sergeants how to use them. It also helps if you have been to a lot of schools, and if you have been in the division for a year, so that you know the detailed procedure for making the system operate. When you know your stuff you can talk directly with the soldier about his chores. One of the last tasks I had in that company was to hold an 0600 class for lieutenants on how to adjust the sights on all of the company's weapons.''

As for gaining the soldiers' confidence by telling them the truth,

Andersen felt the point to be too obvious to merit long discussion. "As a platoon leader, I watched the company commander lie to the troops. If you lie to them or make promises that you know you can't keep, they will soon find out, and you will have lost all real communication with them. They will go their own ways, just as you have, and they will look out for themselves, just as you have.

"Showing soldiers that you are concerned about them means more than mouthing platitudes about truth and trust. It goes right down to how they live—getting the barracks painted, building TV and game rooms that they can actually use, finding decent furniture. It is amazing how much material is in the supply system—after you have been turned down on the first three tries. Concern also means things like confiscating stereos from those who won't keep the noise-level down. Soldiers will not complain about this in an era when everyone is supposed to be allowed to do their own thing, but they sure appreciate fewer decibels when authority says that's the way it will be.

"And you have to deal directly with all their personal problems. The First Sergeant assured fair treatment by the creditors and the landlords of the troops living with their wives off-post. The executive officer, Lt. William G. Graves, and I were troubleshooters in those centralized support systems that bypass the chain of command and deal directly with the soldier on his pay, benefits, transfers, dependent allowances, and so forth. Few civilians realize that we are required to run a company without a clerk or typewriter. Rather, everything is centralized at a higher level where the soldier gets lost among people who care nothing about him. You show concern when you open your underground admin center and make sure things are done right.

"You show concern by continuing counselling. Every new soldier talked with the company commander, where my pitch was 'Honesty, Loyalty, and Your Ability to Get Things Done *Right*.' I talked winning and success with them. In the long run, it is success that brings them into the company fold. About the time I left Bravo Company we were beginning to get good soldiers who had just arrived on post, coming to talk with the First Sergeant about getting into the company. Success breeds success.''

THE ROOTS OF COMMAND SUCCESS

I was somewhat disappointed by what Andersen told me. There is nothing new about establishing clear goals, enforcing high standards, developing subordinates, and taking care of people. It appeared that he was fully applying the old rules, whereas others may not have been. When I inquired further, I found that very few commanders were punishing and discharging to the extent that he was, nor were they spending as much time on expressing goals, promoting, refurbishing barracks, and winning competitions. I asked why not.

The answers varied. In some cases, it was a matter of ignorance concerning the tools available to the company commander. Many officers command their first company without having studied command techniques in schools. They learn by observing their predecessors, a practice that guarantees the replication of errors and a sameness of method. Commanders are also subject to the demands of their superiors, who may prefer safer policies than Andersen was pursuing with Bravo Company. He himself anticipated that he might not achieve the same results in his next command if he did not inherit higher authorities who would give give him the trust that he had experienced in the Bramlett-Dawkins-Louisell chain at Fort Campbell.

There are also philosophical disagreements over Andersen's techniques. To some, his system of rewards for the winners is excessive—"You shouldn't reward a man for merely doing his duty." And, as for his culling the inept and uncontrollable out of the Army, one view says, "The officer's job is to make soldiers out of the raw material he gets." As the Army became understrength for lack of volunteers, more leaders in Washington advocated doing away with expeditious discharges in favor of "rehabilitation transfers" for a fresh start in another unit. While Andersen did use the "rehab transfer" on a few occasions, he thought that any greater use would be no more than shuffling his misfits on to another commander.

In addition to these impediments of ignorance, habit, command influence from above, and certain philosophical hang-ups, there was one great deterrent to action—a lack of conviction that building cohesive units is important. There was little in the officer learning system that said, "Your first task is to build a strong unit, a unit in which the individual willingly subordinates himself to the good of the whole." Rather, emphasis was on the primacy of individual soldiers—their rights, their training, their duty hours and free time, their education, their benefits, their re-enlistment. A company commander could earn maximum efficiency ratings by focusing on these requirements, but never weld together the total environment that begets compact fighting units.

We know there are many company commanders who are unconcerned about either the units or the individuals within them. These are marginal officers whose interest is in personal survival throughout a command tour. In this way, they can stay in the Army and provide for their families. They will comply with directives from above, cut corners to achieve moderate successes, and sacrifice the interests of soldiers and organizations to avoid being numbered among the 50% of captains who never become lieutenant colonels. They are not hostile to setting goals, demanding high standards, developing subordinates, and taking care of their soldiers; they are just indifferent to these propositions when the time comes to take action. Their minds are elsewhere.

How do we explain why Andersen and others like him place importance on creating good military organizations? Perhaps it was better training and education that gave Andersen the tools and the confidence to dare new things. He had nearly 40 weeks of military training before being commissioned. Subsequently, he completed Airborne, Ranger, Air Assault, Jungle Warfare, and a variety of lesser professional schools. While this training is indeed a factor, others with the same experience did not exploit it as Andersen did. Technical competence is essential, but not enough.

Some argue that combat units are built by professionally oriented officers as part of pride in their profession. Andersen, however, rarely used the words "profession" in discussing his philosophy of command. Nor did he ever refer to his many hours in leadership classes, which he found invariably boring and of little practical or theoretical value.

The fact that Andersen was a bachelor meant that he could spend 16 hours each day in the company, devoting weekends to writing the administrative documentation that accompanies every personnel action. It also meant that he had less financial strain when he gave his personal funds to support special projects in the company. But, other bachelors chose not to follow the same path.

In the final analysis, Andersen was a highly successful commander because he cared. He cared for his people, his unit, and his principles. He cared for his higher commanders. He cared enough for himself to demand that he himself live up to the exceptional standards he had set for all. He sensed the wisdom of S.L.A. Marshall, who had written shortly after World War II:

it might be well to speak of the importance of enthusiasm, kindness, courtesy, and justice, which are the safeguards of honor and the tokens of mutual respect between man and man. This last there must be if men are to go forward together, prosper in the bonds of mutual service, and experience a common felicity in the relationship between the leader and the led.

—S.L.A. Marshall, *Men Against Fire,* 200.

ANDERSEN'S KOREAN POSTSCRIPT

Within weeks of leaving Campbell, Andersen assumed command of another Bravo Company, this time in an infantry battalion of the Second Infantry Division in Korea. "It was a very different world," he said, "with people on a one-year tour in an alien land. Personnel instability dominated every action, since a 120% annual turnover meant that even the leaders spent 3 months breaking in, 6 months doing their jobs, and 3 months refitting for the next one. Time was too short for processing many bad soldiers out of the Army. The living conditions were primitive, with rusted quonset huts open to the winter wind, and the latrines and showers a long walk through the snow. Morale seemed to be based on the soldier's other life with his 'yobo' in the 'ville,' and the commander's personnel problems were highlighted by blackmarketing, venereal disease, and alcoholism.

"Training was oriented on one month of the year, which the company spent patrolling and guarding the demilitarized zone facing North Korea. Here life was orderly and predictable, with the troops' attention riveted on the live ammunition they were carrying and on the odds of being shot at by a fanatical enemy. But, at home in Camp Hovey, it was a world of disorganization and suspicion, where unpredictable changes in orders from above kept the company in continual chaos. A company commander's training program could be wiped out overnight by a colonel ordering the unit off on an 80-mile road march, whose purpose was little more than to seek a few headlines in the post newspaper. As soon as I had launched 20 soldiers into a high school diploma study program—6 weeks, 6 hours a day, with an imported civilian instructor—the whole battalion was sent off on an impromptu training exercise that was poorly planned and executed. A company could be summarily dispatched on a four-day weekend exercise, without warning, simply because someone had asked for an aggressor detail and the battalion commander felt that 'they don't have anything else to do anyway.' With this calloused view of soldiers, I suspect we were woefully underprepared for a real firefight.

"I was expected to lead the company in any direction my bosses wanted, regardless of my assessment of what was good for the unit or the development of the soldiers. I resisted the worst of the broken promises, changing priorities, and foolish waste of soldier time. Yet, I was highly rated in my performance, probably because they knew I was right. At the end of 6 months, however, I had not established a climate of trust in the company, and I was not upset by abrupt orders to move to Division Headquarters.

"There I learned that companies who had better battalion commanders were in considerably better shape. But we all faced the effects of stepped-up readiness 'requirements' which, in effect, called for twice as many inspections and test exercises as are used in stateside divisions. In some outfits, this meant that the lieutenants spent most of their time in the motor pools, supply and commo rooms, and personnel records centers, doing the work of enlisted specialists who, in such high turnover, were generally not prepared to set and maintain the high standards demanded by inspection checklists. One battalion commander called this the 'trivialization of officership.' Before I left Korea, the numbers of inspections and tests began to decline, but they still detracted from the training of units and the development of soldiers.''

Andersen's commands occurred in a snapshot of time in the life of the Army—years in which manpower was short and of questionable quality, support from the public and the Carter administration was low, and the armed forces were carrying out more missions than was justified by their resources. He found both good and not so good commanders above him. He saw the reality of what he was doing and shaped his command policies to fit the immediate situation at hand.

Yet, his experiences showed how little common understanding there was about the role and powers of the company commander, as well as the proper relationship between senior and junior commanders in peacetime organizations. Too often one concludes that good or bad companies are the products of good or bad battalion commanders, rather than good or bad company commanders.

Captain Bill Andersen's command experience underscores the extensive education and training necessary for exemplary decision making and leadership at the lower levels of command. But it was not necessary for him to wait until six or seven years of service to perform in an exemplary manner, as long as he had learned his lessons well by the third. He was not aware, however, that Army policy has (for several decades) sought greater control and efficiency at the lowest level, and has achieved this at the expense of the authority and responsibility of junior commanders. Andersen learned how commanders at battalion and higher levels can waste the excellence that some of these junior officers possess, forcing officers to become obedient followers rather than the aggressive, creative individuals needed in combat.

It appears that the rule should be: Company commanders should have much the same authority, responsibility, independence of action, and institutional support in peacetime as they can expect in combat in wartime. Senior commanders, out of concern for the professional development of their juniors, try to make this rule live.

LEARNING ABOUT COMPANY LEADERSHIP

Books for studying the phenomena of peacetime company command are few in number and almost never entertaining. Exceptions are Anton Myrer's *Once An Eagle,* with its chapter on the between-the-wars American Army, and John Master's *Bugles and A Tiger,* an instructive memoir about learning to be an officer in the British Indian Army in the thirties. Aside from these, most of the literature is turgid and sometimes hard to understand, designed more for classroom dissection than for personal inspiration.

The formal study of leadership by cadets and junior officers has travelled a rocky road. This started in the twenties, when suggestions about teaching leadership were rebuffed by generals pronouncing that leadership could be learned only by experience in a unit. By the end of World War II, however, the disciplines of the behavioral sciences had developed sufficiently for Army Chief of Staff Dwight Eisenhower to direct that ROTC and Military Academy cadets take courses in psychology and leadership. The service schools, especially the Infantry School at Fort Benning, followed suit, and by the 1960's the responsibility for leadership learning seemed to have passed from the individual to the institutions. Textbooks describing the traits and techniques of the good leader (Hays, *Taking Command,* 1967) gave way to analyses of how leaders interact with groups in varying situations (USMA, *Leadership in Organizations,* 1983).

The authors of these texts drew extensively upon the research and writing of behavioral scientists like Maslow, Stogdill, Steers, and Porter. Stogdill, for example, had analyzed more than 100 studies on leadership. He found a consensus that successful leaders, regardless of their occupations, tended to carry out common functions. These are summarized in Chart 3. In a similar vein, Steers and Porter, in *Motivation and Work Behavior,* increased our understanding of why some leaders—like Captain Andersen—can summon up great sources of energy in pursuing their leadership goals. In 1981, James H. Buck and Lawrence J. Korb edited *Military Leadership,* a collection of essays updating some of the conceptual theory. In that book, David R. Segal challenged assumptions that organizational management theories based on individual self-interest can be applied to military organizations. And Lewis Sorley, in "The Leader as Practicing Manager," described the tendency of modern generals to overload units with missions without providing the necessary resources for accomplishment.

As the research and writing of this literature became more complex, and more devoted to a specialized vocabulary, Army authorities began to seek a return to more familiar territory, at least for the indoctrination of new officers in leadership principles and techniques. Proponency for this instruction was moved to Fort Leavenworth, where a 1982 version of Field Manual 22-100, *Military Leadership*, signalled a return to more prescription of what a leader must know, be, and do. Historical example was used extensively in illustrating the message, and the character traits of the good leader were again emphasized.

Once the novice combat arms officer leaves his basic schools, his best leadership laboratory is in the troop unit, dealing with soldier problems every day. This assumes, however, that the inquiring soldier keeps his mind alive with infusions of ideas outside his daily existence. Self-study of leadership in these circumstances can take several directions, such as in the how-to "cookbooks," the combat monographs, or the mind-stretching accounts of leaders in politics or industry.

The military "cookbooks" provide checklists and summaries of how one should organize his thinking or actions in the vast number of leadership situations he faces. *The Armed Forces Officer* is the most familiar of these, written by S.L.A. Marshall shortly after World War II, and more recently issued as Department of Defense publication GEN-36. Maj. Gen. Aubrey "Red" Newman's *Follow Me* is a 1982 collection of his "The Forward Edge" columns that appear in service

Chart 3

Leader Behavior

Stogdill's Patterns of Behavior of the Effective Leader:

- applies pressure for productive output

- exhibits foresight and ability to predict

- maintains closely knit organization and resolves intermember conflicts

- maintains cordial relations with superiors and has influence with them

- actively exercises the leadership role and does not surrender leadership to others

- allows followers scope for initiative, decision, and action

- clearly defines own role and lets followers know what is expected

- uses persuasion and argument effectively, partly by exhibiting strong convictions

- tolerates uncertainty and postponement without anxiety or upset

- reconciles conflicting organizational demands and reduces disorder within the system

- speaks and acts as chief representative of the group in all significant situations

—Stogdill, *A Handbook of Leadership,* 143

---

magazines, devoted to the leadership problems of junior officers in "the old Army." General Bruce C. Clarke's *Guidelines for the Leader and the Commander* was published in 1963 from the articles, speeches, and directives he used to lead the World War II generation of officers. Finally, Lt. Gen. Arthur S. "Ace" Collins' *Common Sense Training* has, since 1978, been praised by many junior officers as most helpful for developing their techniques in leading and training in the troop units.

Company grade officers who have not served in combat can fight through World War II in Europe in *Company Commander,* whose author, Charles B. MacDonald, began the book:

because I was a captain, my lot was easier sometimes than that of Joe Private. . . . But when my lot was easier physically it might be harder

mentally, because I knew Joe Private, and Joe Private First Class, and
Joe Sergeant and Joe Lieutenant, and I could not suppress my love and
admiration for them.

—MacDonald, *Company Commander,*v.

In 1985, MacDonald wrote that he had "long waited for someone
to do for the platoon leader what I had tried thirty-eight years ago to
do for the company commander." He finally found this in James R.
McDonough's *Platoon Leader,* a graphic memoir of his command of
an understrength platoon in an isolated outpost in Vietnam, suffering
endless casualties, but never flinching from their mission. MacDonald
praised the book: "It will become a classic."

During and after World War II, S.L.A. Marshall interviewed soldiers
and wrote *Men Against Fire,* describing how men can be conditioned
to act as a cohesive unit under the stress of battle. "All things being
equal," he said, "the tactical unity of men working together in combat
will be in ratio of their knowledge and sympathetic understanding of
each other. Lacking these things, though they be well-trained soldiers,
they are not likely to adhere . . ." (150). Marshall raised fundamental
questions, still germane today, of why soldiers fail to fire their guns
in battle, and how the lack of moral leadership can destroy the effectiveness of fighting organizations.

Of course, these ideas are not entirely new. During the First World
War, American officers discovered and translated Ardant duPicq, a
French colonel who wrote his *Battle Studies* on ancient and modern
warfare in the mid-1800's, before being killed in action in the 1870
Franco-Prussian War. DuPicq argued that moral force is the most
powerful strength in combat, and that the study of warfare must start
with the study of the human beings who wage it. He found fault with
those who said that new technology made weapons more instrumental
in bringing victory than those who manned them. Rather, duPicq argued:

Since the invention of fire arms, the musket, rifle, cannon, the distances
of mutual aid and support are increased between the various arms. The
more men think themselves isolated, the more need they have of high
morale. . . . We are brought by dispersion to the need of cohesion
greater than before.

—duPicq, *Battle Studies,* 21.

In the duPicq tradition, American officers were schooled further in the European battle experience through lectures given by German Captain Adolph von Schell, at the Infantry School. His book, *Battle Leadership,* was reprinted in 1982 by the U.S. Marine Corps Association. For similar reminiscences of British junior officers, Americans in the 1930s read Robert Graves' *Goodbye To All That,* and Siegfried Sassoon's *Memoirs of An Infantry Officer*—books written with a grace of style and insight that has disappeared from military literature. Today we seek factual data and analysis, seen best in W. Darryl Hendersen's *Cohesion: The Human Element in Combat* and Larry Ingraham's picture of the 1980s peacetime soldier, *The Boys in the Barracks.*

Given the availability of books on company leadership, can we be certain that reading them will make one a better leader? I asked Captain Jack H. Cage who, like Andersen, commanded a company early in his career. He did so well in this and other leadership assignments that he was placed on a promotion list to major in his tenth year of service. Meanwhile, he had earned a Columbia University PhD in psychology, and had taught leadership at the Military Academy. Cage applauded the Army's continuing research in the phenomena of military leadership. He also confirmed that cadets and officers should undergo training in the theoretical bases of leadership, from texts such as *Leadership in Organizations* and/or the Hersey and Blanchard *Management of Organizational Behavior.*

Cage also knew, however, that few junior officers state that this formal study has been of any great value to them in their work. Was this because theory was based on models and aggregates of human beings, too faceless for easy identification by the individual? Or was the jargon too impenetrable? Cage reminded me of the danger of judging the validity of theory by its applicability to any single case. He also suggested that leadership, like tactics, requires practice and feedback before success can be assured.

He then proposed that the junior officer follow up theoretical study in the schools with the best available practical literature, designed to be of use to the man on the job. He liked most of the historical and biographical accounts I suggested. But he was particularly attracted to two recent books, popular for their simplicity and practicality. The first was Colonel Dandridge "Mike" Malone's *Small Unit Leadership: A Common Sense Approach,* a superb book for teaching leadership within troop units. Malone accurately sensed that leadership is not an isolated entity, but part of a process used to achieve clearly stated and finite goals. Cage's second recommendation was Blanchard and Zigarmis' *Leadership and The One Minute Manager,* the third of the popular "one minute management" series. It was designed to increase one's managerial effectiveness by selecting the leadership style most suitable to the situation at hand, whether it be directing, coaching, supporting, or delegating. Overall, Captain Cage thought that most commanders could profit from the reading of books, assuming that they integrated this learning with the practical experience of the workplace.

Does the officer at the company level have time to read? Captain Bill Andersen thought that those who wanted to found the time to do so. Most of his reading time was consumed in the daily exercise of finding the right manual, circular, regulation, pamphlet, or policy statement—then analyzing it, broadcasting it, and assuring compliance with it. General Louisell prescribed his reading in tactics, including The Department of the Army publication *Soviet Army Operations,* Rommel's *Attacks,* and Swinton's *The Defence of Duffer's Drift.*

By 2130 on a "normal" day, however, Andersen could turn to his own reading, which focused on biographies and the lives of prominent leaders in all walks of life—their characters, motives, significant decisions, speechmaking, how they attained and used power, and how they were influenced by ethical principles and values. Typical was John F. Kennedy's *Profiles in Courage.* At any given time he had three books in progress, finding time on weekends, during air deployments, and in those long waits in field exercises. He used some of the rules learned in speed reading courses, took no notes, rarely underlined, but occasionally copied some passages into his journal. For Andersen, reading was not in conflict with his work; it was a part of it. It made him more effective. It gave him perspective.

Reading About Company Command and Leadership

In answer to the question, "What should I read about company level command and leadership?," those books marked with an asterisk are recommended for priority in personal study or in an officer professional development seminar:

* Kenneth H. Blanchard, Patricia Zigarmi, and Drea Zigarmi, *Leadership and The One Minute Manager*.
* James H. Buck and Lawrence J. Korb, *Military Leadership*.
* General Bruce C. Clarke, *Guidelines for the Leader and the Commander*.
* Arthur S. Collins, *Common Sense Training: A Working Philosophy for Leaders*.
* Ardant duPicq, *Battle Studies*.
  Richard A. Gabriel and Paul L. Savage, *Crisis in Command*.
  Robert Graves, *Goodbye To All That*.
  Samuel H. Hays and William N. Thomas, ed., *Taking Command: The Art and Science of Military Leadership*.
  Wm. Darryl Hendersen, *Cohesion: The Human Element in Combat*.
  Paul Hersey and Kenneth H. Blanchard, *Management of Organizational Behavior: Utilizing Human Resources*.
  Larry H. Ingraham, *The Boys in the Barracks: Observations*.
  John F. Kennedy, *Profiles in Courage*.
* Colonel Dandridge Malone, *Small Unit Leadership: A Common Sense Approach*.
* S.L.A. Marshall, *Men AGainst Fire*.
  A.H. Maslow, *Motivation and Personality*.
* John Masters, *Bugles and a Tiger*.
* Charles B. MacDonald, *Company Commander*.
* James R. McDonough, *Platoon Leader*.
  Charles C. Moskos, Jr., *The American Enlisted Man*.

*   Anton Myrer, *Once An Eagle*.
Maj. Gen. Aubrey Newman, *Follow Me*.
General Matthew B. Ridgway, *Soldier*.
Siegfried Sassoon, *Memoirs of an Infantry Officer*.
Captain Adolph von Schell, *Battle Leadership*.
Richard M. Steers and Lyman W. Porter, *Motivation and Work Behavior*.
Ralph M. Stogdill, *Handbook of Leadership: A Study of Theory and Research*.
*   U.S. Department of the Army, FM 22-100, *Military Leadership*.
U.S. Department of the Army, *Soviet Army Operations*.
*   U.S. Department of Defense, DOD GEN-36, *The Armed Forces Officer*.
U.S. Military Academy, Assoc. in Behavorial Sciences and Leadership, *Leadership in Organizations*, 1983.
U.S. Military Academy, Assoc. in Military Leadership, *A Study of Organizational Leadership*.

CHAPTER FOUR

The Commander as Tactician

Nine-tenths of tactics are certain, and taught in books; but the irrational tenth is like the kingfisher flashing across the pool, and that is the test for generals. It can only be ensured by instinct, sharpened by thought, practicing the stroke so often that at the crisis it is as natural as a reflex.

—T.E. Lawrence, *The Science of Guerrilla Warfare.*

In peacetime, battalion commanders who know little and do not care about tactics may remain unexposed; their subordinates can cover for them in short training tests on restricted terrain. In war, however, their incompetence quickly surfaces, hopefully before doing too much damage.

Fortunately, we also have many peacetime commanders who are excited by the art and science of tactics; they are never forgotten by their subordinates, who later tell how their old commander taught a daily school of tactics, in what he did and said and required them to study. These teachers of tactics are not only commanders in the combat arms, but are also quartermasters and adjutants general. There is something about being a soldier that compels all to dabble in tactical theory and practice, regardless of station and style.

My generation was raised to believe that tactics was the fun part of being a soldier. Our commanders had become master tacticians as they fought through World War II, discovering by experience new techniques and theories for winning battles. We and they probably did not know what classic ''tactics'' was. In our minds it was maneuvering

59

small forces about the landscape in order to surprise and defeat the enemy, with proper attention to supporting fires and communications.

At some point, we were told that the word stemmed from the Greek *taktos,* which refers to things that are ordered or arranged, such as in formations. In later schooling, tactics came to be defined as the planning, training, and control of military formations when engagement by rival forces is imminent or underway. This terminology included grand tactics, minor tactics, and applied tactics.

If strategy was a matter of bringing military force *to* the battlefield to accomplish some policy goal, tactics was a means of applying such force to the enemy *on* the battlefield. It all seemed less grand, however, when we extracted tactical kernels from manuals labelled "The Armored Infantry Battalion" or "The Reconnaissance Platoon," and restated our new knowledge in classes taught by Departments of Tactics at the service schools. But it all came alive again when we maneuvered our small commands at Fort Hood or Fort Benning or on the old Wehrmacht training grounds at Grafenwohr.

Our first visions of the Great Tacticians came from the early lessons of our cadet courses in military history. There were four ancient commanders who defeated enemy forces vastly outnumbering their own; this was considered a good standard for measuring tactical success. *Miltiades* did so at Marathon in 490 B.C. by employing a double envelopment against the Persians. *Epaminondas* of Thebes employed an oblique order (advancing with units in echelon) in his attack on the Spartans at Leuctra in 371 B.C., and defeated 11,000 troops with his 6,000. *Alexander* of Macedon honed the oblique order into a hammer and anvil single envelopment and, in 333–331 B.C., destroyed larger Persian forces at Issus and Gaugamela. A century later, *Hannibal* of Carthage brought the Roman Army into the classic battle of annihilation at Cannae, where his double envelopment slaughtered some 44,000 of the enemy at a loss of 6,700 of his own. The legacy that a great commander should be a great tactician was thus started, a legacy that has been nourished by Frederick, Napoleon, and the modern day architects of Sedan, Tannenberg, Dunkirk, and El Alemein.

Of course, not all practicing tacticians are commanders. Some are operations officers and designers of tactical doctrine. Colonel Robert Doughty, author of *The Evolution of U.S. Army Tactical Doctrine, 1946–76,* was asked to name the ten outstanding tacticians of all time; he selected those whose visions of new possibilities in weaponry and equipment might transform the battlefield. One of his tacticians, Jacques de Guibert, described in the 1780s a new order of line and column, new uses of artillery and cavalry, and a new citizen soldier from an emerging new society. It is said that Napoleon Bonaparte read Guibert's five volumes and developed his own vision from the possibilities. Today we study Napoleon, the consummate tactician, forgetting the source and essentiality of the tactical vision that drove him.

Currently, in the late twentieth century, we are probably most comfortable with the image of Erwin Rommel as a great tactician. Perhaps it is because he made his reputation on the unusual terrain of the North African desert that he could rise above the fate of so many good tacticians whose achievements lie cloaked in the great tide of mass armies, suffocating doctrine, and bureaucratic authorship known as modern warfare. We know of Rommel's victories, plans, and thoughts, and how he maneuvered smaller forces against larger ones partly because he recorded and saved everything. We are surprised to find that his first command of armored forces came late—in 1940 when he took a panzer division into the battle against the French. But he had done his homework in earlier years, and had built a vision of himself as a tactician in writing *Infanterie Greift An,* a critique of tactical concepts in World War I (now published under the title *Attacks*). Today, *The Rommel Papers,* with their commentary by B.H. Liddell Hart, are a monument to the development and flowering of a commanding modern tactician.

Should a great tactician be able to handle all kinds of warfare? This is a very tall order, if one reflects on the demands represented by the following sample of books on tactical operations:

- From World War I: Alistair Horne's *The Price of Glory: Verdun 1916*

- From Vietnam: S.L.A. Marshall *Battles in the Monsoon* or John Albright, et al, *Seven Firefights in Vietnam,* or Dave R. Palmer, *The Summons of the Trumpet*

- From the Middle East: Chaim Herzog, *The Arab-Israeli Wars* or Lt. Gen. Saad El Shazly, *The Crossing of the Suez,* or Avrahan Adan, *On the Banks of the Suez*

- From the different services, as seen in the biographies of air General Henry "Hap" Arnold, *Global Mission,* and the naval tactician

Admiral Raymond Spruance, as told in Thomas B. Buell's *The Quiet Warrior*

- From low-intensity warfare: Walter Laquer, *Guerrilla,* or Charles R. Simpson, *Inside the Green Berets*.

Perhaps no one person can master all kinds of tactics in one lifetime.

Are there any attributes common to all tacticians, regardless of era, locale, or mode of operations? It has been written that all tacticians take risks and that many are gamblers. C.S. Forester portrayed Horatio Hornblower as a whist player; books have been written on the similarity between poker and war. In 1924, Liddell Hart compared the tactician to the boxer [later published in *Thoughts on War*]:

> watch a boxing-match, and a few minutes' reflection will enable us to grasp the principles that govern all tactics. Every action is seen to fall into one of three main categories—guarding, hitting, or moving. Here, then, are the elements of combat, whether in war or pugilism.
>
> What is the man's object in guarding himself? To obtain security against being knocked out by his opponent before he can get in his own blow. . . . so a military body pushes out a detachment to secure it against surprise or interference with its plan. . . .
>
> When contact is made with a definite enemy resistance, the phase of guarding changes into that of hitting. As in boxing, we lead with one fist, to fix the opponent's attention and create an opening for a knock-out blow—which may be delivered by either fist. . . . Here in a nutshell is the key formula of all tactics, great or small—that of *fixing combined with decisive maneuver,* i.e., while one limb or part of your force, whether an army or a platoon, 'fixes' the enemy, the other part strikes at a vulnerable and exposed point—usually the flank or communications in war just as it is the chin or solar plexus in boxing. If the junior leader will think of his advanced troops and his reserve as his two fists, his understanding and explanation of tactics will be simplified. . . .

—B.H. Liddell Hart, *Thoughts on War,* 274–276.

Liddell Hart further explains that his boxer is constantly moving, in order to create the best conditions for guarding and hitting. It is with this image of the tactician-as-boxer that he develops his thesis of maneuver as the key to victory. Its use is the mark of the superb tactician, regardless of time, place, and culture.

ESSENTIALS OF THE TACTICAL ART

The knowledge and skills of the master tacticians are in four areas: terrain, hardware, doctrine, and a cultivated capacity for creativity. Knowledge of hardware and doctrine must, of course, encompass the essentials of logistics and administration that make the tactical plan come alive.

Terrain

The classic account of how to organize a piece of military terrain is *The Defence of Duffer's Drift,* written by British Captain E.D. Swinton before World War I, and reprinted and parodied by American soldiers of later generations. This succession of bad decisions and calamitous defeats occurred in the dreams of a lieutenant who had not had a lifetime to think about hills and valleys as gun positions, breastworks, fields of fire, and obstacles to an enemy advance. Ultimately, he got it right.

Fortunately for the American Army, terrain analysis became a central theme for cadets and officers before World War II, including topographical drawing, terrain walks, and long hours of map-reading instruction. As a result, many of the biographies of World War II generals refer to conversations with their wives and fellow travelers about military crests and defilade positions, as they motored across the country to some new camp or station. Today, old timers mention with regret the crowding out of these studies from military curricula by more modern requirements.

The helicopter has brought a new and different picture of terrain to the commander, following the World War II use of the light plane to spot things in the enemy rear that Napoleon could never see. We are only a short distance from the commander's constant use of satellite scanning of the battlefield, where foliage and camouflage can be stripped away with new sensing techniques, and where electronic guidance can place missiles on enemy targets despite traditional safeguards of defilade and rear-slope positions.

The new technology transforms the manner in which terrain can be used, but not the essentiality of matching terrain with the capabilities of weapons and formations. The ability to read a map, and to see terrain through its symbols and elevation lines, seems to increase rather than diminish with new technology.

Lt. Col. Wesley Clark uged his officers at Fort Carson to use their junior years in constant study of maps. At first, he said, it is appropriate to ask "Where am I on this map?" But, as their sense of the tactician grew, they should ask of the map: "What is the best way of moving a military force through this terrain?" Not too many years later they should begin to ask this same map: "If I were organizing this terrain for defense by a combined arms team, what geographical features would guide my thinking about the kind of defense I would use?" Clark was talking to combat arms officers, but his three questions would seem pertinent to the thinking of logisticians, administrators, and communicators.

Hardware

Once the first mechanized battalion was created, commanders began telling their lieutenants, "Get thee to the motor pool." For just as long, lieutenants have been saying, "I am wasting away my life in the motor pool." The commanders have been right in wanting lieutenants to learn about the hardware while supervising maintenance. The lieutenants have been right because unthinking commanders would not say how much is enough, thereby allowing lieutenants the time to learn other things. Many of them were indeed wasting their time.

How much knowledge of weapons, vehicles, equipment, and radios is enough? Enough to deploy, fight, and maintain forces, and to troubleshoot their failures. Enough to make them function with all their potential. Clark taught his battalion that a special Man-Machine Bond had to be achieved. "If you are firing your tank downrange and your score is low because your turret malfunctioned, you don't get an alibi rerun if the crew lacked the ability to troubleshoot the problem on the spot. That is what you would have to do in combat if you expect to survive." Clark met the companies each morning in the motor pool. Nearly all training, testing, and personnel actions transpired in the motor pool, until the companies marched back to the barracks at nightfall.

How much is enough? Most of the Army's senior generals whom I knew in the 1970s had reputations as good tacticians, and as hardware men in their junior officer years. Maj. Gen. James C. Smith had emerged from the Sixth Cavalry Regiment in World War II as the quintessential lieutenant-communicator, extracting unheard-of performances from radios by positioning them, calibrating them, throwing antenna wires into trees; he was also a tanker, mechanic, and aviator. Lt. Gen. Arthur G. Trudeau's early experience as an Engineer gave him special expertise in river crossings and amphibious landings at a time when the Army needed them most; he later headed important research and development activities, commanded the 7th Division in Korea, and was the Army ACS for Intelligence. Lt. Gen. Paul F. Gorman became such an expert on the gadgetry of the mechanized infantryman that he could lead the school at Fort Benning into a new mobile era and recast Army doctrine and training for the last quarter of the twentieth century. In the same vein, the careers of generals James Gavin, William Depuy, Hamilton Howze, and Donn Starry had their roots in a special knowledge of and interest in promising pieces of hardware—the helicopter, the parachute, the antitank missile, whatever. In all these cases, "enough" knowledge usually meant "more than anyone else around."

During the Vietnam War, "technical and tactical competence" became the buzzword for the first requirement of all soldiers. Why not just say "tactical competence," since no officer can be a competent tactician without a deep understanding of the technology involved? Because, I was told, too many soldiers misread tactics as a theoretical game in which symbols of units are pushed about on mapboards—a never-never land of dreams, detached from the real battlefield and the fog of war. If this is so, then we need to enshrine the concept of technical competence and, when necessary to get people's attention, mandate tests of Military Qualification Standards to assure minimum levels of competence. The commander, especially in times of Army expansion and wartime mobilization and high turnover of military personnel, is "The Technician of Last Resort" to his unit, who knows where the oil filters are and how far the weapons will fire. The thinking commanders probably learned how to be encyclopedias of technology on their way to becoming shrewd tacticians.

Doctrine

When a new piece of hardware appears on the scene, there ensues a search for a doctrine through which the tacticians will employ it. Doctrine is refined theory — theory by which hardware and terrain are meshed for the accomplishment of some purpose. Not any theory will do, however; it must be accepted by the practitioners and promulgated by a respected authority. Doctrine provides guidance for action by commanders; it facilitates communication between them, in hopes of assuring coordinated efforts on the battlefield. Doctrine becomes the officially approved teaching in military classrooms, and it is the base from which new organization and new equipment are developed. The tactician uses doctrine as the foundation for his creative solution of problems on the battlefield.

In our first days in the Army, we learn how the rifle squad should function; that is doctrine and it is in the manuals which, collectively, are the expression of doctrine for the practitioners. Because it takes years to make a manual, however, the tactician with an eye to the future scours service journals, accounts of equipment tests and training exercises, and even newspaper reports of distant wars—all of which send signals of oncoming changes in doctrine.

Periodically, the American Army overhauls its combat doctrine, announced by the rewriting of *FM 100-5 Operations,* the keystone manual from which all other how-to-fight manuals are derived. The overhaul of the early 1980s was a continuation of that of the mid-1970s, when the lessons of the Arab-Israeli War of 1973 were wedded to the anticipated introduction of some forty new technological systems into the American Army. The 1980's *FM 100-5* was expected to convey to the Army a doctrine of AirLand Battle operations, including striking deep in the enemy rear. This seemed to alter substantially the previous view of conventional battle, and prompted a thorough discussion of tactics, operations, and strategy within the Army, as new field manuals for the Corps, division, brigade, and battalion grew out of the keystone manual. The history of these developments is told in John L. Romjue's *From Active Defense to AirLand Battle: The Development of Army Doctrine, 1973–1982.*

THE TACTICIAN AND THE SCHOOLS

When attending Army schools, officers learn about hardware and doctrine, and they rehearse tactical decisions, using maps and faculty-designed scenarios. Few qualified tacticians emerge solely from this short exposure, but most acquire foundations for future self-study. They become habituated to procedures for analyzing tactical problems, such as the Estimate of the Situation, and learn the value of standard operating procedures and checklists. These tools assure that even the dullest of soldiers travels all the bases while he decides what to do.

Tactics instruction in schools must answer a battery of perennial questions that are inherent in the effort. First is the accusation that the schools are too often out of date. The captain coming from the 101st Airborn Division is shocked to find that the new air assault techniques worked out at Fort Campbell are not yet being taught at Fort Benning. This is the problem in establishing an Army-wide standard for Army-wide instruction, entailing testing, evaluation, and meshing new concepts with the old. Without constant attention, traditionalists can "murderboard" needed changes into extinction, deserving the accusations that dead minds keep dead doctrine alive.

A related charge is that schools tend to lose touch with the reality of combat, as the last war fades into distant memory. Colonel George C. Marshall wrote in 1934:

> There is much evidence to show that officers who have received the best peacetime training available find themselves surprised and confused by the difference between conditions as pictured by map problems and those they encounter in campaign. This is largely because our peacetime training in tactics tends to become increasingly theoretical. In our schools we generally assume that organizations are well-trained and at full strength, that subordinates are competent, that supply arrangements function, that communications work, that orders are carried out. In war, many or all of these conditions may be absent.

This paragraph introduced *Infantry in Battle,* a 400-page anthology of World War I tactical operations, which the Infantry School faculty published in the early thirties under Marshall's direction as Assistant Commandant. Characterized in 1937 by a British quarterly as "the most important military text since 1874," the book is being used at Leavenworth in the 1980's to impart the realism of war to future commanders. (Pogue, *George C. Marshall,* I, Note 24, 397)

Are the schools to blame if their graduates seem guilty of malpractice in their wartime assignments? Colonel Paul F. Gorman thought so upon returning from his second tour in Vietnam, where he commanded an infantry brigade. The problem was one of preventing staff college trained S-3's from defeating search and destroy operations with their Leavenworth goose eggs. When the infantry company commander had received his mission-type order to search an area of so many square kilometers and to destroy any enemy found therein, the S-3 would take him aside to provide added guidance. Where does the commander expect to enter the area and at what time? This then became a line and time of departure on the S-3's map. And where would he go first—this riverline or that hill? Soon the riverline or hill had a goose egg, with a designation as Objective A, at the end of a route of advance, with phase lines and an estimated time of arrival at each. And so on, identifying objectives B through Z, the last being the exit point from the area at sundown. The company was thus sent on a fast roadmarch from A to Z, reporting their progress through checkpoints to the satisfaction of the operations staff. If terrain were to be searched as dictated by discoveries on the ground, or if the enemy were to be pursued and pinned, the Leavenworth habit must be rooted out.

Gorman was subsequently brevetted to brigadier general and given Marshall's old job of running the Infantry School. Chief of Staff Westmoreland directed that he begin to reshape Army instruction for the post-Vietnam era, with an eye to the tactics best suited for the technology of the eighties. In this and his subsequent assignment to Training and Doctrine Command as DCS for Training, Gorman crafted the ideas that were soon to emerge in the redrafting of *FM 100-5, Operations*. The stage was being set for a dramatic revitalizing of the teaching of tactics in the service schools.

The key to this renaissance was the founding of the Combat Studies Institute (CSI) at Fort Leavenworth in 1979. Under the direction of Colonel William A. Stofft, the CSI designed new courses for service schools and ROTC, assuring that future military officers be exposed to the history of a broad variety of battle situations and to a method of battle analysis designed to provide critical thought about tactical employment of military forces.

These programs were concerned with more than merely getting the dull boy to touch all bases. Every officer was grounded in Jessup and Coakley's *Guide to the Study and Use of Military History,* published by the U.S. Army Center of Military History. In most schools, reading requirements were established that introduced student officers to, for example, John Keegan's *The Face of Battle,* an absorbing overview of three great battles in history, as seen by the common soldier. A focus on Korean War tactics required the reading of T.R. Fehrenbach's *This Kind of War,* excellent for portrayals of small unit actions, and S.L.A. Marshall's *The River and the Gauntlet.* For an analysis of tactical development, assignments were made in Robert Doughty's study of American tactical doctrine since 1946, Messinger's *The Blitzkrieg Story,* and in the histories of the branches, such as *A Perspective on Infantry* by John A. English, *Mounted Combat in Vietnam* by Donn A. Starry, and *Air Assault: The Development of Airmobile Warfare* by John A. Galvin.

In the more advanced electives at the Army's colleges, officers began to investigate the important works of the theorists of tactics, such as J.F.C. Fuller's *Armored Warfare,* or Jay Luvaas' *Frederick the Great On The Art of War.* For a better understanding of modern conceptions of maneuver warfare, they chose B.F. Liddell Hart's appeals for "the indirect approach" in his *Thoughts on War* and *Strategy.* At the Combat Studies Institute, the best of the journal articles on tactics were collected into *War and Doctrine,* a readings text for a subcourse of the same name. Its goal was "to create an intellectual environment conducived to innovative thought concerning war and how to wage it and to provide a historical context for the study of tactics."

By 1984, the CSI was producing a stream of new books tailored to the reading needs of the Army officer. Jonathan M. House wrote *Towards Combined Arms Warfare: A Survey of Tactics, Doctrine and Organization in the 20th Century,* a superb synthesis of the development of weapons and their use in the major world powers. In addition, Charles Heller and William Stofft edited a collection of new writings on the initial military engagements of each of the nation's wars in *American First Battles.*

The tactics curriculum at the Command and General Staff College was reshaped in the early eighties, principally by Colonels John Orndorff and Claude Monderson, Directors of the Departments of Tactics and Combat Support, respectively. Maj. Gen. Dave R. Palmer, the Deputy Commandant, assured that the center of gravity of instruction was shifted upward to include divisions and corps, as well as serious study of the operational level of warfare. The blending of tactics, logistics, staff operations, and joint Air Force and Navy considerations, along with leadership and military history—in true team teaching—was the most important curriculum change made in mid-career officer professional development since World War II. It evolved naturally as a new synthesis of ideas and technology, of concepts and doctrine, and paved the way for a more progressive view of battle for an Army. There evolved legitimate interest, serious study, and writing of high quality on the conduct of today's AirLand Battle.

TACTICS AS A SELF-DEVELOPED ART

The difference between those tacticians who have learned by experience and self-study and those who are only school trained became clear to me when I encountered Lt. Gen. William M. Hoge as IXth Corps Commander in Korea in 1951. At a morning briefing, he was told the extent of a major Chinese breakthrough in the center of the peninsula, the IXth Corps zone. The Reds were pouring down the roads and valleys, and after the destruction of so many allied combat units, there was little between the enemy and the Corps Headquarters. He said he would take no action at that time, and went to see for himself.

The evening briefing, with even more dire tidings, elicited the same response from Hoge. The next morning's briefers reflected their mounting concern about getting the vast headquarters on the road; instead, he suggested that a reconnaissance company be placed on a certain hilltop to report what they saw. Throughout the day, refugees and deserters poured by the headquarters, and virtual panic gripped the staff. But he did not flinch, and went to bed.

The next morning, he opened the briefing by reminding us that this was the third day of the Chinese offensive and that they would soon run out of ammunition and food. Meanwhile, they will not have achieved their objective, which is Seoul, far to the west of us. He noted that although they had broken our front, they had to continue diagonally across our sector, since they did not have the flexibility or doctrine to change direction in an attack of this scale. He suggested that the recon company would tell us when it is time for a counterattack into their flanks as they retreat. We now realized that there never was a need to move the headquarters.

The embarrassed staff later discussed the mindset of this rather silent and distant commander, knowing that he had commanded CCB of the 9th Armored, which had so dramatically seized the Remagen bridge over the Rhine on March 7, 1945. Here was a man who believed so strongly in his insights about the enemy that he could predict and take tactical risks that seemed contrary to all the collected wisdom around him. Schools may have set his foundations, but his skill as a tactician came from a commitment to learn while on the job.

In November of 1982, General Richard E. Cavazos published a letter entitled "Officer Training" over his signature as Commanding General of U.S. Army Forces Command. It began:

> Unit performance during training at our installations and the National Training Center and during major exercises shows a distinct shortfall in the tactical skills of many of our officers. While this deficiency is neither new nor peculiar to Forces Command, the ramifications are significant. Our units are not being properly trained and we have officers who are not properly skilled to lead our soldiers in combat. Rapid and imaginative corrective action is required.

He asked his commanders to see that seniors train subordinates throughout the chain of command, "to enhance the combat skills of our officers."

At the 101st Airborne Division, Maj. Gen. Charles W. Bagnal responded with new instructions about officer professional development classes and battle simulation exercises. He also included a reading list of books such as Anton Myrer's *Once an Eagle,* Swinton's *The Defence of Duffer's Drift,* and Rommel's *Attacks,* with the comment that "each officer has the personal responsibility to understand the doctrine and the historical basis that guides the profession of arms."

These letters were important signals to a busy Army that good officers learn continually, rather than just in schools. A year later, some lieutenants at Fort Campbell confessed that they had never heard of the letters nor had they ever been directed to read a book. But others told of battalion commanders who sought out the books, held monthly classes, and even took officers for tactical studies of Civil War battlefields. Some, like Lt. Col. John A. Cope, said the objective was to give officers who had never been in combat an appreciation for its reality; Cope used the descriptions of the effects of artillery detonated in trees, as told by Charles McDonald in *The Huertgen Forest.* If the objective was also a better sense of American combat tactics, they could have used Cornelius Ryan's *A Bridge Too Far,* John Eisenhower's *The Bitter Woods,* and some unit histories, such as Kenneth Koyen's *The Fourth Armored Division.*

If one's objective is to learn about enemy tactics, as Hoge did, reading could include the 1978 U.S. Army manual *Soviet Army Operations,* Parotkin's *The Battle of Kursk,* or David R. Glantz' two masterly volumes on 1945 Soviet battles in Manchuria, known collectively as

August Storm. If the objective is to relearn the tactical lessons handed down by the Great Captains, one should read—or reread—the basic texts, the best of which are the early volumes of *The West Point Military History Series,* edited by Brig. Gen. Thomas E. Griess—especially May, et al, *Ancient and Medieval Warfare;* Britt, et al, *The Dawn of Modern Warfare;* and Britt's *The Wars of Napoleon.*

THE TACTICIAN'S VISION

All these reading objectives remain secondary to that of achieving the special vision that marks the master tacticians, that "acute sense of the possible" which strikes them when they seek to make a decision. Napoleon referred to this as "coup d'oeil," that glance of the eye across a piece of terrain—that glance that brings into focus all the tactician's knowledge and experience, and sets in motion a series of quick decisions concerning how and where to deploy forces.

Instead of talking about a phenomenon of the eye, the Germans refer to "Fingerspitzengefuehl"—that fingertip-feeling or instinctive sense of matching terrain with doctrine and weaponry. One might ask why the Americans have no similar slang for the quick insights of a master tactician. Could it be because there have been so few? Patton wrote about "coup d' oeil," but other American leaders seem strangely silent on the subject.

Those who have coup d'oeil seem to possess a predilection in the mind that occasionally says "now is the time to do what I have been wanting to do." At that point the commander makes a move to the flank, or begins to focus on the possibilities of a certain weapon, or commits a special unit, or moves into a rapid retrograd. Perhaps each idea has been pre-positioned in the commander's mind, having long before said, "I want to try this when the conditions are right."

Experts on creativity say that creativity occurs after the mind has been well honed and stocked with facts and ideas. This storehouse becomes exposed to a question, time goes by, and suddenly the pieces fall into place and a solution appears. Some people, the real geniuses, are more endowed than others, but everyone can improve their creative capacity with training. On this premise, the Stanford University Graduate School of Business (among many) offers a course on creativity in business.

The military service has inherent barriers to the growth of creativity in its commanders. The stifling effects of peacetime bureaucratic politics, the penalties for making decisions unpopular with superiors, and the concentration of power in a rigid chain of command all militate against the creative personality. More important is the strong belief that a battlefield populated with commanders who are creating their own doctrine and organization is a battlefield of total chaos. Army people prefer to talk about innovation rather than creativity, and equate imaginative decision-making with "good common sense." The difficulty, of course, is that the thinking of the master tactician is not at all "common."

An orchestra conductor once described his approach to a score that was created by a great composer like Beethoven, and how he tried to *re-create* it in the most excellent fashion that the composer could have imagined. He said that he and his orchestra members might achieve that excellence in only a very few minutes of a long evening concert. This approximates the work of the tactical commander, who places the reigning doctrine against certain conditions of terrain, enemy, and mission, and tries to re-create what was intended by the doctrinal visionaries in its most excellent form. If creativity is "seeing things in a new way," then it can happen both in creating the book and applying it.

Perhaps this has been better understood by the Germans than by the Americans. In *The Dynamics of Doctrine: The Change in German Tactical Doctrine During The First World War,* Captain Timothy Lupfer described how the Germans closely studied the battlefield experience, induced new principles of doctrine from it, and then turned to the battlefield commanders to bring it into reality. The German professionals retained a great capacity for flexibility in their tactical operations, allowing commanders to re-create on the ground the visions of excellence they had nurtured in their training. Thus, some of the most instructive reading on command is in Heinz Guderian's *Panzer Leader,* Ronald Lewin's *Rommel as Military Commander,* Erich von Manstein's *Lost Victories,* and F.W. von Mellenthin's *Panzer Battles.*

GRAND TACTICS OR OPERATIONS?

Study of the memoirs of World War II's German generals has again raised for Americans the question of whether warfare at the division and corps level is tactics blown up into grand tactics, or whether it is an entirely different phenomena, thought of by the Germans as the operational level of warfare. There is the implication that commanders who have been good tacticians at the battalion level may not necessarily be as good at higher levels unless they learn a new doctrine, a new appreciation for terrain and hardware, and a different outlook on military history.

In 1984, the case for an operational level of warfare, between tactics and strategy, was being made by the military thinkers at Fort Leavenworth to include Colonel Bill Stofft, Colonel Huba Wass de Czege, Lieutenant Colonel L. Don Holder, and Lieutenant Colonels Harold Winton and Douglas Johnson of the School of Advanced Military Studies. They defined the operational level of warfare as "the movement, support, and sequential employment of large military formations in the conduct of military campaigns."

When I asked if large formations meant divisions and corps, they agreed partially, but argued that I should focus more on time and space factors than on the size of the units involved. Stofft suggested that military strategy for a theater of operations might be set at the Army Group level; the campaign plan in turn might be developed into an operational plan at Army level. But it is at corps and division where the plans are turned into fighting operations.

As an example of excellence at the operational level, Stofft cited General J. Lawton Collins' book *Lightning Joe* as "a treatise on the operational level of war," in those sections that concentrate on Collins' command of the U.S. VIIth Corps in Europe in World War II. Hal Winton cited Grant's operations at Vicksburg in 1862–63 as another classic example. In both cases, the ideas of "movement, support, and sequential employment" seemed to take precedence over the more tactical ideas of fighting the battle on the battlefield. A great commander's creativity at both the operational and tactical levels can be studied in Field Marshal William Slim's *Defeat Into Victory,* a gripping story of the loss and reconquest of Burma by allied forces in World War II.

The Germans, French, and Russians, more than the British and Americans, have recognized the uniqueness of the operational level, disposed as they have been on large land masses, facing sizeable enemy ground armies over extended periods of time. In its brief forays into large theaters of war, the American Army may have made more mistakes at the operational level than at the tactical level. Too many senior generals may have allowed their excellence in tactics to blind them to the opportunities for victory in the movement, support, and sequential employment of large forces. They may not have been as psychologically capable of letting the enemy penetrate their zones as deep as 50 miles, as the Germans were on the Russian front, in order to create opportunity for a counterattack into the enemy's flanks and rear.

On the other hand, Collins wrote that he had learned how to deploy the divisions of a corps at Fort Leavenworth in the 1930s. And others note that, by studying Napoleon, Grant, et al, American cadets and officers have always had a good introduction to what we are now calling "operations." Changing the name from Grand Tactics to the Operational Art may help signal to commanders that they need to shift gears mentally as they progress to broader responsibilities. It will be some time before we refer to good senior tacticians as good "operationalists." We might in time, however, find an American expression that distinguishes the commander of great vision, who is known for having "fingerspitzengefuehl" or "coup d'oeil."

Reading For Tacticians

In the following list of books accenting the theory and practice of the tactician's art, those marked with an asterisk have been selected for a balance of readability, scope, and analysis. They are candidates for an officer seminar on "The Commander as Tactician."

Avrahan Adan, *On the Banks of the Suez.*
John Albright, et al, *Seven Firefights in Vietnam.*
Henry H. Arnold, *Global Mission.*
Albert S. Britt, et al, *The Dawn of Modern Warfare.*
* _____ , *The Wars of Napoleon.*
Thomas B. Buell, *The Quiet Warrior.*
* J. Lawton Collins, *Lightning Joe.*
* Robert A. Doughty, *The Evolution of U.S. Army Tactical Doctrine, 1946–1976.*
John S.D. Eisenhower, *The Bitter Woods.*
John A. English, *A Perspective on Infantry.*
T.R. Fehrenbach, *This Kind of War.*
J.F.C. Fuller, *Armored Warfare.*
John A. Galvin, *Air Assault: The Development of Airmobile Warfare.*
David M. Glantz, *August Storm: Soviet Tactical and Operational Combat in Manchuria, 1945.*
Heinz Guderian, *Panzer Leader.*
Russell Gugeler, *Combat Actions in Korea.*
Charles E. Heller and William A. Stofft (eds.) *American First Battles.*
Chaim Herzog, *The Arab-Israeli Wars.*
Alistaire Horne, *The Price of Glory: Verdun 1916.*
Jonathan M. House, *Towards Combined Arms Warfare: A Survey of Tactics, Doctrine, and Organization in the 20th Century.*

John E. Jessup and Robert W. Coakley, *A Guide to the Study and Use of Military History.*

John Keegan, *The Face of Battle: A Study of Agincourt, Waterloo, and The Somme.*

Kenneth Koyen, *The Fourth Armored Division.*

Walter Laquer, *Guerrilla: A Historical and Critical Study.*

Ronald Lewin, *Rommel as Military Commander.*

* B.H. Liddell Hart (ed.) *The Rommel Papers.*

_____ , *Strategy.*

* _____ , *Thoughts on War.*

* Timothy T. Lupfer, *The Dynamics of Doctrine: The Change in German Tactical Doctrine During the First World War.*

Jay Luvaas, *Frederick the Great On The Art of War.*

* Charles B. MacDonald, *The Battle of The Huertgen Forest.*

Charles B. MacDonald and Sidney T. Mathews, *Three Battles: Arnaville, Altuzzo, and Schmidt.*

Erich von Manstein, *Lost Victories.*

S.L.A. Marshall, *Battles in the Monsoon.*

* _____ , *The River and The Gauntlet.*

* Elmer C. May, et al, *Ancient and Medieval Warfare.*

* F.W. von Mellenthin, *Panzer Battles.*

Charles Messenger, *The Blitzkrieg Story.*

Anton Myrer, *Once An Eagle.*

Dave R. Palmer, *The Summons of The Trumpet.*

Ivan Parotkin, *The Battle of Kursk.*

George S. Patton, Jr., *War As I Knew It.*

Forrest C. Pogue, *George C. Marshall. I.*

John L. Romjue, *From Active Defense to AirLand Battle: The Development of Army Doctrine, 1973–1982.*

* Erwin Rommel, *Attacks.*

Cornelius Ryan, *A Bridge Too Far.*

Saad El Shazly, *The Crossing of The Suez.*

Charles M. Simpson, *Inside The Green Berets.*

* Field Marshall Sir William Slim, *Defeat Into Victory.*

General Donn A. Starry, *Mounted Combat in Vietnam.*

* E.D. Swinton, *The Defence of Duffer's Drift.*

U.S. Army Combat Studies Institute, *War and Doctrine.*

* U.S. Army Infantry School, *Infantry in Battle.*

* U.S. Department of the Army, Field Manual 100-5, *Operations.*

U.S. Department of the Army, *Soviet Army Operations.*

CHAPTER FIVE

The Commander as Warrior

I see many soldiers;
could I but see as many warriors!

—F.W. Nietzsche, *Thus Spake Zarathustra,* i, 10.

In the reconstruction of the American Army after the Vietnam War, senior commanders called for an officer corps that would exemplify "the warrior spirit." They intended to build an Army leadership that would endure any hardship, stand firm in battle, and "close with the enemy and destroy him" in steadfast pursuit of victory.

Unfortunately for their purposes, military literature and history contained little that distinguished between their idea of the warrior and the barbarian, sacking cities and piling skulls on the steppes of Asia. More recently, the word had been used callously by editorial writers and movie makers to depict bloodthirsty mercenaries exploiting Asia and Africa for power and profit or, still worse, knuckle-dragging Neanderthals with titles like Conan, The Destroyer.

George S. Patton was one of the few, if not the only, modern American general to write of the soldier as bred and trained to have "the warrior soul":

> Success in war lurks invisible in that vitalizing spark, intangible, yet as evident as lightning—the warrior's soul. . . . It is the cold glitter of the attacker's eyes, not the point of the questing bayonet, that breaks

the line. It is the fierce determination of the driver to close with the enemy, not the mechanical perfection of the tank, that conquers the trench. It is the cataclysmic ecstacy of conflict in the flier, not the perfection of the machine gun, which drops the enemy in flames. Yet volumes are devoted to arms; only pages to inspiration.

—G.S. Patton, "Success in War," *Infantry Journal,* January, 1931, 23.

Unfortunately, Patton never fleshed out this romantic view of what the soldier should be into policies and programs for generating such a warrior. The pragmatists and behaviorists considered it quaint, but useless.

Who else, then, do we have in mind when extolling the warrior spirit? Perhaps Alexander the Great, who bore many battle wounds and scars by the time of his death at thirty-three. But we know that the warrior's hand-to-hand combat with the enemy faded as the musket, the machine gun, and now the long-range missile separate the combatants.

Or perhaps we want to emulate the British subaltern of the nineteenth century, when Army lore dictated that the officer's purpose in wartime was to show the troops how to die. And, until late in World War I, there were those who strode the parapets in defiance of death, until it was proven that a swagger stick was little comfort in an era of shrapnel munitions. Today we no longer use the phrase "show the men how to die"; rather, we say, "show them how to live."

Or, perhaps, some see the ultimate warrior in the Japanese samurai, represented best by Miyamoto Masushi, who hacked sixty opponents to death before the age of thirty. Shortly before his death in 1645 he wrote his philosophy of "The Way of the Sword," advising:

Be intent solely upon killing the enemy. . . . If the enemy is less skillful than yourself, if his rhythm is disorganized, or he has fallen into evasive or retreating attitudes, we must crush him straightaway, with no concern for his presence and without allowing him space for breath. It is essential to crush him all at once.

—Masushi, *A Book of Five Rings,* 80 and 87.

The republishing of his book in the 1980's was marked by passages commending this philosophy to American corporate managers in the modern era of business ethics.

And so we pursue the ethos of the warrior, from the annals of Ghenghis Khan and the Mangoday, as in Charol's *The Mongol Empire,* to the strange biography of Yasuo Kuwahara, one of the Emperor's suicide pilots, in *Kamakaze.*

We, therefore, expect today's commanders to emulate Alexander's courage in facing dangers, not only in the close combat of war in the cities and jungles, but against constant artillery and aerial bombardment. We expect commanders to have the aggressive spirit of the samurai, and to expose themselves to the same dangers that beset their troops, in the tradition of the British officer corps. That these expectations continue to be met is proven in Bruce Jacob's *Heroes of the Army: The Medal of Honor and Its Winners* and in Malcolm McConnell's 1985 true tale of an Air Force pilot, *Into the Mouth of the Cat: The Story of Lance Sijan, Hero of Vietnam.*

Despite the antique, distorted, and romantic views of the warrior, there are still many valid demands on commanders to prepare themselves for warrior roles. They must be personally courageous to function usefully in the hazardous and chaotic conditions of the battlefield. They cannot allow fatigue to cloud their minds. They must get their troops to fight. They must wage violence competently. And they must win on the battlefield, regardless of obstacles, bad luck, and the incompetence of others. Courage, aggressive leadership, skillful war-waging, and winning are all hallmarks of the warrior, in the modern era as well as in the antique past.

THE SOLDIER'S COURAGE

Ben O'Donnell was a bosun's mate in the South Pacific, and was bayoneted in hand-to-hand combat on Guadalcanal. Thereafter, he was the best plumber on the Jersey Shore. When I asked O'Donnell, "What are the values of the Good Soldier?" he immediately replied, "Freedom and courage. Freedom is what makes us fight, and courage keeps us from running away."

Those who would command soldiers in combat understand both why men fight and why they do not run away. The wellsprings of the warrior spirit come not only from the aggressive, animalistic depths of man's nature, but also from his most philosophical and idealistic yearnings. An informed dedication to the cause of freedom's triumph over tyranny inspires men to march to the sound of the guns, as much today as in any age of patriotic fervor. Courage, like bravery, has been the first requirement of the soldier since the most primitive days. Good commanders talk easily and thoughtfully on these matters.

Aristotle, not believing in extremes, placed courage in the median between cowardice and rashness. Plato thought that courage ranked with wisdom, temperance (moderation), and justice among the four chief attributes of the virtuous man. Military biographers tend to mark physical courage as the one human trait whose absence causes soldiers to fail; in few other professions is physical courage such a mandatory qualification.

For the commander, it matters little whether courage is inborn or acquired. More important is how it ebbs and flows in the soldier's breast. Henry, the hero in Stephen Crane's *The Red Badge of Courage,* was initially favorably disposed towards heroism, then ran in terror from the battlefield at Chancellorsville, was restored by a quirk of fate, and finally led the charge into the enemy lines and was decorated.

This fluctuation between Aristotle's cowardice and rashness was also observed by a British battalion surgeon in World War I; in later years he recorded these observations in *The Anatomy of Courage* when, as Lord Moran, he was the personal physician of Winston Churchill. Moran developed the thesis that courage is achieved by constant struggle against fear, and that when fear takes over, the soldier's habits and conduct undergo major change. He concluded that men enter battle with a certain capacity for courage, which is drained through months of great danger, especially that afforded by constant artillery and air bombardment. He wrote:

> There seems to be four degrees of courage and four orders of men measured by that standard. Men who did not feel fear; men who felt fear but did not show it; men who felt fear and showed it but did their job; men who felt fear, showed it and shirked. . . . The story of modern war is concerned with the striving of men, eroded by fear, to maintain a precarious footing on the upper rungs of that ladder.

> —Lord Moran, *The Anatomy of Courage,* 3.

Within Moran's philosophy one finds an explanation of studies of American World War II soldiers, who tended to lose their effectiveness after going beyond six months of unrelieved combat. There are few today who agree with the nineteenth century belief—perhaps harbored by Stephen Crane—that courage is increased by continued exposure to fighting and the hazards of the battlefield.

Based on his interviews with American soldiers in World War II, S.L.A. Marshall wrote that "the seeds of panic are always present in troops so long as they are in the midst of physical danger." He reaffirmed J.F.C. Fuller's belief that, in an attack, half the men on a firing line are in terror and the other half unnerved. Marshall concluded that, as a result of fear, barely twenty-five percent of infantry soldiers in World War II fired their weapons when engaged closely with the enemy. (Marshall, *Men Against Fire,* 70 and 146)

In *This Kind of War: A Study in Unpreparedness,* T.H. Fehrenbach describes how the Americans went to war in 1950 with an Army that was untrained, ill-disciplined, and barely organized. He cites repeated examples of soldiers not firing their weapons, retreating and leaving positions undefended, being physically unfit for their tasks. Courage could not thrive under these conditions, except among the sprinkling of brave junior officers and noncommissioned officers, who suffered inordinate casualties to save the situation until discipline and training could be energized.

Instilling the aggressive spirit in one's troop is always necessary, but never entirely successful. Some commanders have preached "kill or be killed", and some have embellished that edict with a command to "make their arms and legs fly." Bosun's mate O'Donnell admired Admiral "Bull" Halsey for emblazoning a South Pacific harborside with an enormous lighted sign: "Kill Japs. Kill Japs. Kill all the Lousy Bastards." Whether these displays produced any greater warrior spirit or simply induced more atrocities is debatable. But they left no doubt as to the commander's attitude towards their mission.

But there have been just as many commanders who have found such appeals to homicide and thuggery unprofessional and unnecessary in the waging of battle. They see soldiers as being different from ordinary killers. In 1962, Marine commandant General David M. Shoup testified before a Senate committee that the Marine Corps should not be teaching hate, but should be teaching men how to defend themselves and their country. He himself had earned the Congressional Medal of Honor at Tarawa, in Halsey's South Pacific.

Professionals like Shoup operate with neither hatred nor fear of the enemy. They are little stirred by emotional appeals, and may even be reserved about broadcasting their deep sense of patriotism. They respect their enemy, with whom they often identify themselves—in the same predicaments, with the same goals, fighting the same weather and terrain, hoping for the same breaks. Across the battlefield they see other individuals very much like themselves.

We learn through reading and experience that the best guarantee of the warrior spirit, of courage in combat, is the discipline imposed on soldiers' minds by a tight military organization, where orders are given and obeyed in a crisp fashion and the habit of carrying out tasks and functions is ingrained through constant practice. We have also found the soldier's courage to be fortified by confidence in his ability to complete those chores that are at the very limit of his capability. We introduce him into the Army with tests of his courage, such as the slide-for-life, the obstacle course, pugil sticks, bayonet drill, and live fire exercises. We also know that soldiers in good physical condition can more easily fight off the fatigue that breeds fear and defeatism. This knowledge has been pulled together by Anthony Kellett in *Combat Motivation: The Behavior of Soldiers in Combat,* an excellent book completed in 1980 for the Canadian Department of Defence.

When the British Imperial Army was at its zenith, it assured that the soldier was immersed in one unit, for his entire seven-year enlistment, in a life that never varied in its regimen, whether in Edinburgh or the Punjab. John Baynes describes such a unit, the Second Scottish Rifles, in his *Morale: A Study of Men and Courage*—how the soldiers were recruited, trained, disciplined and, finally, how they fought to the death at Neuve Chapelle in 1915. Once recruited into the battalion, soldiers rarely ventured into civilian life, but were under the constant surveillance of their officers and NCO's in training and sports; they were professional bachelors. Discipline was constant and had two objectives:

> . . . to insure that the soldier does not give way in times of great danger to his natural instinct for self-preservation, but carries out his orders even though they may lead to his own death.

> . . . to keep order within an Army itself, so that it may be easily moved and controlled, and so that it should not abuse its power.

> —John Baynes, *Morale,* 180.

Although many of these soldiers may not have had a mental age beyond ten years, they were fierce in their loyalty, their patriotism, and their dedication to die when necessary.

Every age, every army, has had its warrior legions—the Spartan hoplites, the Italian condottieri, the French Foreign Legion. More recently, the warrior soul has been kept alive in the Gurkha battalions described by John Masters in *Bugles and a Tiger* and the close-knit bands of Jean Lartegy's *The Centurions*. The common denominators of their trade are competence and courage, fostered by discipline and camaraderie.

The American Army has always had a problem dealing with units considered to be "elite," in that they were held together by special personnel policies. Even the large parachute organizations have had difficulty in keeping their special identity—even their berets. General Ridgway explains in *Soldier* how such units can be misused. The Special Forces units, where morale and skills most exemplify the warrior spirit in the American Army, have had a harrowing battle with survival, as Charles R. Simpson describes in *Inside The Green Berets*.

Fehrenbach's *This Kind of War* is a reminder of what happens when we fail to take action on what we know to be right. Since the Korean War, Army readiness programs have helped to assure that minimum levels of obedience, job training, and physical fitness are part of the soldier's daily life. But the Army has been less appreciative of the injunctions to keep trained soldiers together in cohesive fighting units. DuPicq wrote over a century ago:

> A wise organization insures that the personnel of combat groups changes as little as possible, so that comrades in peacetime maneuvers shall be comrades in war. From living together and obeying the same chiefs, from commanding the same men, from sharing fatigue and rest, from cooperation among men who quickly understand each other in the execution of warlike movements, may be bred brotherhood, professional knowledge, sentiment, above all, unity. . . . Unity and confidence cannot be improvised. They alone can create the mutual trust, that feeling of force which gives courage and daring.
>
> —Ardant duPicq, *Battle Studies*,96.

It remains a mystery why the senior leadership of the American Army took so long after World War II to understand the message written by duPicq and S.L.A. Marshall, and to be sufficiently shocked by the lessons of Korean War unpreparedness. They hesitated to adopt personnel policies that would train soldiers and keep them together in cohesive units. It was not until the 1980's that Army commanders began to fashion COHORT (cohesion, readiness, training) units and some form of regimental system—old Army practices that had died in the World War II individual replacement system. Whether these tentative commitments will survive in the face of efficiency experts, budget cutters, and sly personnel slot-fillers, remains to be seen.

The ability of future commanders to instill the warrior spirit in soldiers, and to fortify their courage to withstand the rigors of combat, depends on whether the soldiers will fight as trained teams and squads and companies—on the first day of battle.

THE WARRIOR'S DILEMMA

Fortunately for the citizen, but unfortunately for the warrior, there are as many demands for restraining him as there are for unleashing him on the enemy. From the start, Western man has shown concern that his wars be justified morally, and that his warriors know and observe certain rules concerning the extent of violence they perpetrate.

The rationale for constraining today's American commander grew from Christian dogma, although the Bible itself does not prohibit resort to war. From the early and medieval church emerged criteria by which princes and soldiers should judge and be judged on whether they are morally justified in going to war (*jus ad bellum*), and whether they are just in waging the war (*jus in bello*).

Recent statements of these criteria have appeared in such American military manuals as "The War Convention," and are summarized in Chart 4.

Only very senior commanders become involved in the national strategy that establishes the first three of these conditions. Any commander, however, can be held personally responsible for the prevalence of the last three conditions.

Observance of these rules have varied throughout the ages with the ebb and flow of man's hatreds, religious convictions, and war treasuries. Great Captains of eighteenth century Europe maneuvered and postured, hoping to avoid the destruction of armies too expensive to replace. In the nineteenth century, the code of the gentleman so

Chart 4

The Traditional Conditions of a Just War

- The war must be declared as a last resort after all peaceful means of settling the dispute have been exhausted.

- The war must be declared by lawful authority.

- The war must be declared for a just cause. For example: To repel serious injury.

- The war should employ just means. There should be no intentional, direct destruction of innocents.

- There must be reasonable expectation of the success of the war.

- The principle of proportionality must be observed. The violence must not produce a greater evil than that which it seeks to correct.

---

infused Lord Wellington that, in fighting Napoleon in Spain, he would often forego the destruction of civilian centers. "Gentlemen do not bombard cities," he said. We fought the Civil War in the same vein; Sherman's attacks on the civil populace were thought to be outrageous. And, in World War I, the German hope of reducing Paris with Big Bertha was viewed as equally heinous.

World War II, however, opened with the blitzkrieg, whose philosophy and implementation called for a smashing *attaque brusque* by well-prepared and disciplined forces, whose very violence was aimed at terrorizing their foes into submission. It closed with the massive obliteration of civilian centers, first using the twentieth century inventions of carpet bombing and fire storms and, finally, the atomic bomb.

The deaths of 20 million soldiers and citizens in the first World War did not restrain humankind from killing twice that number in World War II. But in the half-century that preceded and encompassed those tragedies, the political leaders of the warring nations wrote laws by which they could court-martial and execute military commanders if killing became too indiscriminate and destruction too wanton.

The Hague Convention of 1907, the Geneva Agreements of 1949, the Protocols of 1977, and their companion codociles of international law enlarged on the traditional moral prohibitions against: 1. causing disproportionate harm, destruction, or suffering; 2. destroying or seizing property for other than military necessity; and 3. attacking defenseless peoples, hospitals, and churches. Written into law were the requirements that commanders treat prisoners and civilians humanely and disobey orders that would criminally violate the law of war.

American commanders have responded to these restraints in keeping with the national popular will. As they undertook the crusade against Hitler and Tojo, they received very little criticism for their treatment of noncombatants who were being bombed, and they exterminated the enemy according to a policy of unconditional surrender. Patton wrote in his diary on April 15, 1943: "War is very simple, direct and ruthless. It takes a simple, direct, and ruthless man to wage war." (Blumenson, *The Patton Papers,* II, 221)

The public lapse into total war to defeat Hitler and the Japanese was but a lapse. Thereafter, their misgivings about destruction and civilian casualties reappeared in the Korean War, and appeared full swing in the public demand that Vietnam cities not be bombed, and that even the smallest hamlet be warned before being fired upon.

American soldiers committed atrocities in the Vietnam War, well documented in Seymour Hersh's *My Lai: A Report on the Massacre and Its Aftermath.* Despite the publicity of these tragedies, analysts such as Guenther Lewy in *America in Vietnam* have argued persuasively that American military commanders lived up to the standards required by the War Convention and the rules of engagement set by American authorities. In *Law, Soldiers, and Combat,* Peter Karstens reported that, where crimes occurred against individuals and small groups, these could be attributed to temporary extremes of rage built up in men who had seen their friends killed—or to the inadequate personality traits of unstable individuals, particularly those most beset by ethnocentricity or other forms of fear and hatred. Poor leadership and illegal orders were occasionally evident. But atrocities were not ordered to be committed for reasons of military necessity.

In the aftermath of the Vietnam War, some argued that this war was lost—or at least put in jeopardy—because of excessive restraints on commanders. In 1975, Senator Barry Goldwater told Congress that the bondage of international agreements had become so excessive that senior military chiefs doubted whether a successful war effort could be pursued. A contrary view held that written, legal restraints were not that important in the eyes of most commanders. It was said, that they acted more on their own sense of moral right and wrong, ignoring written rules where they seriously impeded operations, while seeking some sense of moral proportionality in a war with an enemy whose stated tactics were to terrorize the civilian population into submission.

The attempt to circumscribe the warrior with a bureaucratic network of rules, inspections, and punishments continues, mostly on the part of those who have never commanded in battle, nor ever expect to. These words are useful in military education and training as statements of what is expected of wartime commanders. But to expect civil or military leaders in wartime to order commanders to jeopardize the lives of their troops in order to protect civilian populations and property is to underestimate what American public opinion will demand in wartime. In the final analysis, the military commander himself will find those moral edges of policy that seem to fit his situation. Good commanders will establish an appropriate ethical stance; bad commanders will reveal their ignorance, insensitivity, or general incompetence. Higher authority will be required to answer to the press and in civil proceedings why it empowered those inadequate commanders.

The Army teaches its novices that violation of the law of land warfare is militarily inefficient. That is, to torture prisoners is only to invite the enemy to do the same, or to cause the civilian populace to suffer needlessly, only to harden their resistance and make peace more difficult to achieve. These practical arguments may make more sense to the professional warrior than moral or legal arguments. We know that Rommel disobeyed Hitler's order to take no allied prisoners in the African desert. We do not know whether he did this from moral conviction, from an internal obedience to a higher fundamental law, or from a pragmatic concern for his own soldiers. But we do know that he had the power and the will, as a military commander, to exert enormous moral force within his domain, and he did.

A commander's effort to maintain a decent moral climate in wartime is beset with pernicious attacks on that effort. Rage, sadism, and lunacy occur. But there is also the awful recognition that good men can be brought to participate in the degradation. In *A Rumor of War* Marine Lieutenant Phillip Caputo wrote that, after he and his troops arrived in Vietnam, their morals wore away as did the bluing on their rifles. He watched his men degenerate into thuggery, burning the wrong villages and killing the wrong people. Although he was exonerated in his courtmartial, he wrote that, yes, he and other good men had indeed become bad men.

James R. McDonough suggests in *Platoon Leader* that he had been well trained for his moral trials in Vietnam. His platoon was in contact every day with villagers who aided his platoon as much as they did the savage Vietcong. He tried to retain some sense of humanity and to uphold the laws of land warfare. He saw his primary role as a warrior—to press the fight to the enemy, and to work desperately to give his troops the best chance of survival. His story is a classic one, because in the murky waters of wartime morality, even the best of commanders discover that they have erred—given the wrong order, made the wrong assumptions, authorized the wrong activities. They admit their error to themselves, make restitution where they can, and move on to the next decision.

Their model could well be Ulysses S. Grant as a commander in the Mexican and Civil Wars, where he often weighed the moral conse-quences of his actions. In his *Personal Memoirs* he wrote that, as a lieutenant in Mexico in 1848, he had believed that war was declared and waged unjustly by the United States government. (I, 53) Nor was he ever convinced of the justness of the Civil War, in which he became the senior commander of Union forces. These views did not paralyze him into inactivity, however, because he felt he had obligations other than to protest the unjustness of the war—obligations to use his training to end the war as soon as possible, with terms befitting a lasting peace. He made regulations on the just treatment of civilians and prisoners; he respected the Southern enemy; he insisted on the most liberal of peace terms at Appomatox. He brought victory to the Union forces using a form of total war as yet unknown. As a cadet, he had been trained in moral philosophy and the law of war as it existed at that time. Today he is both blamed and praised for his moral judgments in wartime. He was able to cope well with the warrior's dilemma.

ON BECOMING WARRIORS

Some men and women are born into this warrior role, but most have to learn it. As Crane and Caputo point out, many fail these tests of courageous character. But while Caputo's moral convictions were erod-ing with the bluing on his rifle, Commander (later Vice Admiral) James Bond Stockdale, USN, found his moral convictions growing stronger in a seven-year incarceration as a prisoner in North Vietnam. He had

On Becoming Warriors

To fulfill their roles as warriors, future commanders equip themselves for four challenges:

- to be courageous in the face of great adversity.
- to insure that their soldiers will fight with courage and aggressiveness.
- to wage war with the violence necessary to achieve victory.
- to restrain their use of that violence, in order to meet required standards of legality, morality, and justice.

---

suffered from a broken shoulder and knee during bailout from his Jayhawk, after which he was subjected to beatings, leg irons, solitary confinement, and torture. To avoid being used in television propaganda, he cut his scalp and pounded his face with a wooden stool.

Stockdale later said that, to keep his sanity, he concentrated on the philosophy course he had taken as a Stanford University graduate student, and recalled particularly the Book of Job in the Bible as well as the essential teachings of the Greek philosopher Epictetus. He valued the Stoic view that life is unfair, and that to feel sorry for oneself is pointless. Nor, was it useful to hate the enemy, or to dwell on arguments over good versus evil or mind versus body. It was only necessary to keep from giving in to the enemy and, in the long run, to defeat him.

Not surprising, Admiral Stockdale later advocated that all military people study philosophy and history, to undergird the *character* necessary if one is to eliminate the flaws in personality that can subvert the warrior spirit when it is needed. He credited his own real-life experience in competitive sports and tough military training with shaping that character.

Behavioral scientists have toyed with the contention that only certain kinds of personalities are capable of the courage and aggressiveness needed by good military commanders. They look for signs of competitiveness, love of power, toughness in dealing with other people, and a generally extroverted nature. They continue to explore the relationship between the male hormone (in both men and women) and aggressive actions and attitudes. Some argue that psychological testing can assess a commander's capacity as a warrior, and can possibly provide grounds for eliminating those of pacifist or irresponsible persuasion. Perhaps they reconfirm a traditional assumption that some warriors come by their reputations early, while most have to work for them, pushing themselves into continuing choices of right and wrong conduct, accustoming themselves to the consequences of daring, fear, and "being out on a limb." Command comes easier for those who began life with an aggressive outlook in sandbox play, and grew up comfortable with the role of pressuring others to produce results, to move faster, to adhere to a party line, or to give the last full measure of their devotion.

In the wake of World War II, J. Glenn Gray, a teacher of philosophy, wrote a remarkable book on how men go about waging war; appropriately, he entitled it *The Warriors*. Gray suggested that the soldier's basic credo is that life is not the highest good to which one can aspire — he will often sacrifice himself for others, or even for an ideal. While this willingness to sacrifice is nourished by battlefield comradeship, it finds its roots in a rather casual attitude towards death and dying. This seems rather shocking in a society that goes to no end to prolong life beyond realistic expectation. For years, General Creighton Abrams, senior commander in Vietnam and then Chief of Staff of the Army, seemed to remain unheard by reporters and colleagues when he said repeatedly, "There are many things in life worse than dying." His was the warrior's credo.

Gray's journal notes from World War II also described "the enduring appeals of battle" that possess many soldiers and sustain them through their worst fears. He suggested that in all ages, and still today, there are three secret attractions of war. First, there is the delight in seeing spectacle — "the lust of the eye," in witnessing the greatest drama that man stages. Then, there is the delight in comradeship, when the commitment of men to each other in a deadly bond becomes the highest spiritual attainment of their lives. And then, there is the delight in sheer, wanton destruction, which illuminates the dark side of our lives.

General Omar Bradley is said to have told General Patton in World War II, "I do this job because I am trained to do it; you do it because you love it." Both were carrying on as they had been trained to carry on. But is it possible that both actually found pleasure in war? Robert E. Lee confessed to Longstreet that it is well that war is so terrible — we would grow too fond of it. Homer wrote of ancient times, "Men grow tired of sleep, love, singing, and dancing, sooner than of war." Louis XIV mused on his deathbed that he had loved war *too well*.

Warriors on their horses, "grunts" in their choppers, keep alive a legacy of fascination with the very horror of battle, where killing is sanctioned and death offers a merciful end to suffering. And, in the aftermath, especially of Vietnam, men can go insane with the remembrance of their misdeeds, and in their appalling confessions that they loved that carnage, and forever endure a haunting guilt with the ghosts of the past whose deaths they did not share. Gray saw this as the endless legacy of *The Warriors*.

Of the four challenges to the warrior, the most difficult may be that of restraining the violence that lies within his realm of responsibility. Since war must be *measured* violence if it is to be effective in the age of mass destruction, the warrior himself must be the measurer. The application of force and the control of force are the same action, viewed from two different perspectives in the mind of a single commander. We may think of senior generals as "measurers of violence," issuing their Rules of Engagement from their higher headquarters and answering directly to civilian authority. In reality, however, it is the Lieutenant Caputos, from their first days on the battlefield, who must make daily judgments about limiting fire on noncombatants, when it might mean casualties to their own troops. How do lieutenants prepare themselves for such decisions?

On November 26, 1950, Lieutenant Nye's tank platoon was ranged along the Changchun River near Kunu-ri, North Korea. The cold morning sun reflected off the white stone beach, strewn with a hundred Chinese bodies where .50-caliber machine guns had dropped them during a futile attempt at a night crossing. Suddenly, the platoon sergeant ordered Open Fire on a line of white-clad civilians walking down the opposite bank to the river's edge. Lieutenant Nye immediately ordered a cease fire. The sergeant had experience with the enemy, forcing civilians to hide mortar rounds under their clothing and then marching them into forward areas; he therefore protested in anger as the platoon stopped firing. Only a few civilians had fallen.

The showdown between the lieutenant and the sergeant was averted, though, when shrapnel from an incoming mortar round forced the sergeant to be evacuated with wounds. Two tank commanders later told the lieutenant that the platoon was glad to get the cease fire order; the sergeant, they said, had a reputation for being "trigger happy" and was somewhat a sadist in the treatment of his own men. Besides, there could not have been a mortar position on that bank of the river; it was clearly visible with binoculars, as were the civilians who were carrying nothing.

The lieutenant was not so certain, however, that he had not endangered his troops, as the sergeant had contended. And he noticed later that he had some difficulty maintaining fire discipline among the infantry troops attached to him. Perhaps they were confused about the real nature of the enemy and what were and were not proper targets. In a firefight the next day, the lieutenant added his .50-caliber to the other machine guns firing at shadowy figures running across the fields at maximum range, and the troops cheered when he dropped one.

Junior commanders do not have time to consider carefully many of their combat decisions, especially those involving moral dilemmas between two very good principles, such as protect-your-troops versus protect-noncombatants. Military courses in Ethics and Professionalism teach a lengthy process of reasoning one's way through moral dilemmas. But the decisions of junior commanders reflect less of what they have been taught as soldiers and more of the moral characters they brought with them into the Army from their teachers, parents, and childhood environments. This is the personal character that Aristotle referred to as a habit, spawned by years of choosing between right and wrong. That character is best fortified by returning to the same wellsprings from which the original was fashioned.

The Inquiring Soldier often turns, as Admiral Stockdale did, to the discourse of the university and the church, rather than to the military classroom, to establish the foundations for sound moral judgments in military situations. Those who have not had a formal course in philosophy may find basic texts such as John Hosper's *Human Conduct* or William K. Frankena's *Ethics* quite useful. Given this framework, the officer can begin to research the specific concerns of the warrior, such as courage and duty. After exhausting the short reading list mentioned at the end of this chapter, the best source becomes *The Great Ideas,* a two-volume *Syntopicon of the Great Books of the Western World,* which analyzes 102 ideas like courage, and cites readings on the subject from the world's best literature. For example, here we learn that Freud saw courage, not as a product of deciding that something is more precious than life, but more as an intuitive impulse that amounts to "nothing can happen to me." (*The Great Ideas,* I, 257)

Aristotle suggested that a sense of justice was just as important as a sense of courage in the makeup of a worthy military commander. We expect today's commanders to be concerned about the justice that will be achieved for their nation and people if they win the "just" wars that they fight. In leading their troops, commanders are expected to be just in their distribution of tasks and rewards throughout their command, thus achieving morale and cohesion. The nature and importance of justice are best described in the last chapter of Mortimer J. Adler's *Six Great Ideas*.

When listing justice, wisdom, and courage as premier virtues, Plato also counselled that men can undo their best intentions unless they allow temperance to govern their thoughts and actions. This "Greek mean" was more than advice to do all things in moderation. He taught that undue pride and the temptation towards arrogance were potential in all humans. They were a special threat to men given power over other men, and the resulting *hubris* often accounted for the failure of powerful commanders.

In a book entitled *From The Jaws of Victory,* Charles Fair describes how some good soldiers went wrong because of hubris. These men often struck the pose of the warrior and thus blinded soldiers and citizens alike as they lived out their fantasies. Such a warrior was George Armstrong Custer, who exhibited all of his character flaws long before he led his command to the massacre at the Little Big Horn. He was a poor cadet, swaggering in defiance of authority and, at the time of his commissioning, he was under court martial charges for assaulting another cadet. His bravery in the Civil War was as real as his politicking, and by the age of twenty-three he was a brigadier general. But in postwar Fort Riley, Kansas, he treated his men ruthlessly, had soldiers shot for desertion, went AWOL to visit his wife, and was finally court martialed again in 1867 and suspended without pay for a year. Reinstated, he subsequently led his 7th Cavalry in the reckless charge against the Sioux and the Cheyennes, who slaughtered more than 200 good troopers and Custer himself. A recent retelling of the story of this foolishly brave, and therefore criminally deficient, warrior is Evan S. Connell's *Son of the Morning Star.*

The seeds of hubris are present in the makeup of any man or woman who would undertake the powerful role of warrior in today's society, with its mass armies and weapons of great lethality. With the granting of such power, the warrior meets the challenges of being courageous, instilling courage in others, and waging war with the violence necessary to win. Success will come to him, however, only to the degree that his aggressiveness and daring produce results that provide justice for the people and the institutions he represents.

Reading For Warriors

Asterisks indicate a combination of books suited for an officer seminar on "The Commander as Warrior."

    Mortimer J. Adler, *Six Great Ideas*.
*   John Baynes, *Morale: A Study of Men and Courage*.
*   Phillip Caputo, *A Rumor of War*.
    Michael Charol, *The Mongol Empire: It's Rise and Legacy*.
*   Evan S. Connell, *Son of The Morning Star*.
*   Stephen Crane, *The Red Badge of Courage*.
    Epictetus, *The Discourses*.
    Charles Fair, *From the Jaws of Victory*.
*   T.R. Fehrenbach, *This Kind of War*.
    William K. Frankena, *Ethics*.
    Ulysses S. Grant, *Personal Memoirs*.
*   J. Glenn Gray, *The Warriors*.
    The Great Ideas. A Syntopicon of Great Books of the Western World.
    Seymour M. Hersh, *My Lai 4: A Report on the Massacre and Its Aftermath*.
    Paul Horgan, *A Distant Trumpet*.
    John Hospers, *Human Conduct: Problems of Ethics*.
    Michael Howard, *Restraints on War: Studies in the Limitations of Armed Conflict*.
    Bruce Jacobs, *Heroes of the Army: The Medal of Honor and Its Winners*.
*   Peter Karsten, *Law, Soldiers, and Combat*.
*   Anthony Kellett, *Combat Motivation: The Behavior of Soldiers in Battle*.
    Yasuo Kuwahara and Gordon T. Allred, *Kamakaze*.
*   Jean Larteguy, *The Centurions*.
*   Guenter Lewy, *America in Vietnam*.

* S.L.A. Marshall, *Men Against Fire*.
* John Masters, *Bugles and a Tiger*.
  Malcolm McConnell, *Into the Mouth of the Cat: The Story of Lance Sijan, Hero of Vietnam*.
* James R. McDonough, *Platoon Leader*.
* Lord Moran, *The Anatomy of Courage*.
* Miyamoto Musashi, *A Book of Five Rings*.
* Colonel Ardant du Picq, *Battle Studies*.
  John Rawls, *A Theory of Justice*.
  Michael Walzer, *Just and Unjust Wars*.

CHAPTER SIX

The Commander as Moral Arbiter

A few honest men are better than numbers. If you choose godly, honest men to be captains of horse, honest men will follow them.

—Oliver Cromwell, *Reorganization of the Army,* 1645.

Every military organization has its moral arbiters—those people who, by their words and deeds, set the standard for moral conduct. One would hope that the senior commander is the chief moral arbiter in a unit, although this is not always the case.

Not long ago, during a hot summer month in South Korea, an infantry company commander was required to put 120 armed patrols into the demilitarized zone facing North Korea. They carried live ammunition and were expected to intercept infiltrators from the North. One morning, the S-3 (the battalion operations officer, a major) told the company commander that higher headquarters had ordered an additional patrol that afternoon; they were to make repairs on an engineer installation in the zone. The company commander pointed out the reasons why this would violate policies about mandatory briefings for the men, equipment checks, soldier rest requirements, and so forth. The argument became heated and was taken to the battalion commander. After extensive huddling with the S-3, he announced that the company would not have to send out the additional patrol. Instead, higher headquarters would simply be told that it had been done. Faced with the prospect of a conspiracy to lie, the company commander suggested an alternative:

to cancel one of the scheduled patrols, and substitute this special one—same manpower, different mission. Everyone liked this solution, and the plot to lie to higher headquarters was aborted.

This incident raises the question of why some commanders reach their positions so ill-prepared to be moral arbiters for their people. While the issue could be dismissed as being of minor importance, the impact on the organization's moral climate could be major if repeated in the commander's other decisions. In contrast is the commander who establishes a climate that encourages junior officers to make the right moral decisions.

This was the case in a stateside unit, where Lieutenant Jim Riffe was given command of an anti-tank platoon of TOW missiles (Tube-launched, Optically-sighted, Wire-guided). This platoon had been known for having every member qualified as an expert gunner on the TOW. But when Lieutenant Riffe took them to the range and had them evaluated by three sergeants from other units, he found that he had only a handful of experts, the rest having difficulty even in bringing the weapon's simulator under control.

Lieut. Riffe brought all the falsified readiness reports to the battalion commander, who had them destroyed. A new firing program eventually produced an accurate record—it again met the required standards for readiness reports. The moral arbiters had done the right thing. The only question remaining was why the previous platoon leader had felt impelled to falsify the records in the first place.

This story raises some familiar questions. How can commanders be sure that their people are telling them the truth? What accounts for two lieutenants, faced with the same responsibilities, pursuing honest and dishonest means to reach the same results? What consequences do false readiness reports have on the combat effectiveness of a unit? How does a commander prepare to cope with the problems of dishonesty in his command?

In the days of a small Regular Army, a senior commander might assume that the bulk of his officers had been prepared at home and in college to live by some code of honor. This assumption withered in the large standing Army that came with World War II and the Cold War. Those years of change were also marked by increasing deception and fraud in the Officer Corps, as standards more closely resembled those of society in general, rather than those of an elite, professional corps. As a result, frustrated senior commanders often felt impelled to march their lieutenants to the post theater and lecture them on the evils of lying, cheating, stealing, and other moral lapses. Afterwards, the seniors usually felt better, but the lieutenants muttered about having heard it all before.

The lieutenants usually had a point. The often-repeated lectures argued that military people should be honest with each other, since lying caused casualties and failed missions. They heard Story One, where the lieutenant reported that he was at Position A when he was really at Position B; the battalion commander called fire on B and obliterated the platoon. In Story Two, the lieutenant reported taking his patrol to Hill X and finding no enemy, while in actuality he had stopped short and came home without checking it; the battalion commander launched an advance that was snuffed out by an enemy attack from Hill X. Story Three described the lieutenant who reported that his weapons had all been checked; yet in the combat showdown they failed to fire and the unit was lost. As such stories were repeated, they became less and less connected with the wartime experiences that might have produced them. Therefore, they became less believable, since history does not often record military action in such detail.

Additional theoretical arguments for soldier honesty began to appear as the post-World War II generation wrote new texts for classes on modern military morality, politics, and professionalism. Story Four: Congress and the press discover that an Air Force general sent false reports about bombing in Cambodia; they conclude that the necessary public trust in the military's words and deeds has been damaged. This results in fewer funds and manpower for the military.

Story Five: A Colonel makes it known, as he moves from assignment to assignment, that he expects a silver tray or a set of golf clubs as a farewell gift to his wife from the officers of his command; this violates the code of ethics of the profession he espouses. Story Six: A company commander promises his troops weekend passes in return for their extra efforts in preparing for a command inspection; afterwards, he denies having made such promises, thereby forfeiting his ability to lead them in the future. As mandatory classes in Ethics and Professionalism increased in the seventies, arguments for building trust with the public and with subordinates joined arguments for combat efficiency in lectures to military novices.

Outside the classroom, and in quiet conversations, one hears soldiers give different reasons why they should be honest. Some will say, "I want my kids to think of me as an honest man.'' Or, ''I want my friends and colleagues to feel that I am trustworthy and won't let them down.'' Occasionally, one will say, ''I took an oath,'' or, ''God says that I should not bear false witness,'' or simply, ''It is wrong to lie.'' These are reflections of the inner being. They are expressions of the moral character that the soldier brings to his work from his earliest days. They are manifestations of the ''bag of virtues'' which, according to Plato, constitutes the worthy man—virtues such as truthfulness, benevolence, and trustworthiness. When soldiers choose to act in accordance with these virtues, they are doing so to uphold their reputations in the eyes of family, friends, colleagues, their ''public,'' their religion, and their own esteem.

To some degree, military lying and cheating comes from people who have never developed the character traits that will engender good conduct. Such men and women fail to develop a trust for institutions, a high regard for the dignity of others, a sense of justice, or any of the other foundations of good character. It is such people about whom Viscount Morley wrote, ''No man can climb beyond the limitations of his own character.''

More often, however, the military problem is one of coaxing people of good character to act in accordance with their beliefs when operating in situations of great pressure and very little moral support. Behavioral researchers attest that everyone lies or cheats at one time or another, sometimes because they disagree over what constitutes lying or deception, sometimes because they are caught in dilemmas of conflicting loyalties, sometimes because they are trapped into no alternative. This is an environmental or situational viewpoint, which suggests not only that men and women have good character training in their youths, but also that they need favorable conditions in order to exercise those characters. Does the reality of military life meet these expectations?

THE POWERS EXPERIENCE

Few soldiers have had to cope with Army lying more than did Robert D. Powers. He chaired a cadet honor committee, was commissioned in infantry in 1972, and then commanded platoons and a company in a stateside airborne organization. Later, during his command of a company in the Alabama National Guard, I asked him whether he thought that liars and cheaters were people of bad character, or merely average people caught up in bad environments.

"In my five years on active duty," he said, "I learned that most new officers have good enough character, but some must learn how to translate that into real world terms. We had an excellent lieutenant who took pride in a set of stereo headphones that he had liberated from an Air Force C-130 used in jumping. I suggested that this misappropriation of government property set a bad example for the company; he replied that he had paid for the headset with a case of C-rations taken from the Army. I told him that the company commander should decide the issue, but the CO vacillated until I asked to take the matter to the battalion commander. A few days later, the lieutenant told me that he had returned the headphones, and knew that I was right when I had first brought it up. Until then, he had known that he would never steal, but had never been required to admit that his borrowing and misappropriation was, in fact, stealing.

"Most soldiers," said Powers, "are honest by nature and want to do the right thing. This is especially true of young noncoms who like to tell it like it is, and to hell with the consequences. When good noncoms and junior officers begin shading the truth, it is rarely for their own benefit but, rather, for the good of their troops or unit. Unfortunately, they are often misguided about what is 'good.'

"Keeping honest men honest is only one part of coping with lying in the military. Apparently, every major organization has a handful of officers and NCO's who, with malice aforethought, put themselves on a course of lying, false reporting, and other deceptions. My first experience with them was when I was assigned to a special weapons platoon in a headquarters company. This was a notorious collection of AWOL's, drug abusers, brawlers, and a few men with armed robbery on their records. When the next one went absent without leave, I sought a heavy punishment as a multiple offender. But then I found out that his previous offenses had not been recorded; falsified morning reports of "All Present" had been sent forward. Later, the culprit had been punished illegally through the platoon sergeant's work details.

"These false reports were widely used, and gave the company a favorable statistical record on discipline. The company commander had a variety of excuses for this deception; sometimes he blamed the first sergeant's bad paperwork, and other times he said he had found out about it too late to do anything. After I made enough noise, I was able to get my AWOL's reported as they happened, but I don't think the policy changed much for the rest of the company.

"Once I became company executive officer, the scope of the deception became more apparent. When we changed company commanders, I had to review the readiness reports in personnel, training, and maintenance. Weapons firing records had been falsified to inflate statistics on qualifications; in some cases, men were credited with range firing on days when they were not even present for duty. The new company commander and I took the evidence to the battalion S-3 and worked out a plan for range time, ammunition, and schedules that allowed for new qualification. The battalion commander directed that accurate readiness reports replace the old ones as new statistics were developed. Much the same situation appeared in the maintenance and personnel reporting; we even discovered a man being listed present-for-duty when he had been in the Indiana state penitentiary for the past two years.

"You cannot blame this deception on the education and training of that company commander, because he had a Regular Army commission from one of the best colleges with a working honor system. But he lacked competence—he was very nervous, talking and chain-smoking all the time. He was afraid of the men and terrified of the battalion commander. He had a wife and children, and was a candidate for the post-Vietnam forced reduction in the Officer Corps. He was generally disorganized and did not control the events that always engulfed him. To him, lying and cheating was a matter of survival.

"He might have done better under a different battalion commander; ours used intimidation and fear to get what he wanted. He was the kind who refused to let the troops go on pass until a quota of fund-raising tickets for the Youth Activities Center had been bought; our captain would finally pay for the tickets out of his own pocket rather than face up to the men or the commander. The other three company commanders were more successful than my boss. They ran better companies, stood up to the battalion commander, and probably did not feel a need to falsify records. They had the competence and energy to keep up with the game without a lot of deception.

"To some degree, Army policies forced all of us into deception. Company commanders were required to verify all sorts of things with their signatures, things they could not possibly know about. I had to verify that every soldier who left my command had turned in his post library card. Was I supposed to walk each departing soldier to the library? When I had my company on summer training assignments at other Army posts, I signed statements saying my equipment stored at home base was ready for combat. There was really no way for me to inspect it; I was forced to believe the word of my XO who was back there.

"We were all co-opted into writing inflated efficiency reports. We knew that a bad report could be used in Washington to kill the career of a mediocre officer who did an average job; a less than perfect report could permanently put the career of a superb officer to rest. I have had no qualms about rating people higher than the system told me to; for good men, it was a matter of survival. Some of us were enraged when we learned that this captain was given a maximum efficiency report for his command of that company—*after* higher authorities were made aware of his falsification of reports. He was placed on an advanced promotion list to major not long afterwards."

Powers was not surprised to find that this falsification of records went unpunished. "I learned over time that you get little support from above as you expose wrongdoing. They either see a threat to the image of the organization, or perhaps they have been swept into similar practices in the past. In many situations, the wrongdoing cannot be substantiated legally, to the degree that punishment will be carried out. In the case of my company commander, however, we have to wonder if we face a career of working for and competing with men who will do such things and never be held accountable.

"One cannot cope with deception, however, unless he has the courage to investigate it when he encounters it. It is easy to jump to the conclusion that the deception is deliberate and the product of a malicious mind. Very often, however, the problem is one of monumental ignorance rather than deception. We had a company commander who turned in 400 blankets, and a week later blindly signed a statement saying that he still had them; it took us months to straighten out that case of 'reverse deception.' Although in this case, carelessness acted against the individual, there are many situations in which the government is being bilked through the sheer stupidity of its personnel. In such instances, it is foolish to go on a crusade to stamp out deception and evil-mindedness.

"Deliberate deception, however, must be exposed if it is to be corrected. The culprits are usually experienced and know how to protect themselves. If it comes to a showdown between you and them, you will need a pretty good reputation at the higher levels for military competence and good judgment. You will also need to have done your

homework on the administrative and legal details. The experienced
liar and cheater will back away from a power base equal to his, and
will perhaps desist in the practices. Again, higher authority may either
press the issue against him or let him off the hook.

"In coping with lying and other forms of deception, you should
always ask, 'What are the alternatives?' It often appears that soldiers
think they are trapped in a box that can only be broken open by a little
deception when, in fact, they are just too lazy or to uninformed to
discover honest ways of moving forward that would redound to their
credit. Once, when I was in charge of a battalion social fund, I was
told to write a check for an obviously unauthorized expenditure. When
I refused, the battalion commander signed the check himself. Later,
the Inspector General's annual audit of the fund censored the commander and ordered him to restore the money from his own pocket. All
the while, there were alternative ways of getting the money, but the
CO had no intention of pursuing them."

Powers said that, when asked to sign something that he was unsure
about, he frequently used the alternative of attaching a letter that
explained the conditions under which he signed it, or why he did not.
Occasionally, he would get a telephone call asking for further clarifi-
cation, but he was never ordered to reverse his position. "The higher
headquarters realize that they are vulnerable to a charge of setting up
requirements that cannot be met without some deception, and they are
not going to go to the mat to defend their directives. Most people will
tell you privately that they are glad someone raised the issue.

"Anyone who wants to cope with lying in the Army should listen
carefully to the kinds of excuses used by liars to justify their misdeeds.
Of course, their first line of defense is that they are doing it for 'the
good of the men,' or 'the welfare of the unit.' Sometimes, this is
merely a cover-up for their own laziness or incompetence, and the
men and unit would be better off if they simply improved things so
lying would not be 'necessary.' On the other hand, a deceptive act
may bring some good to the unit and troops, but the perpetrator fails
to calculate the long-range impact on the trust of the troops, who are
quick to observe, 'If he lies to them, when will he start lying to me?'"

I asked Powers if we could explore the difference between lies that
were not justified by the excuses and alibis given for them, and the
lies and deceptions that might be validly justified, such as those forced
on the soldier by the efficiency report and verification systems. We

agreed that lying to the enemy in order to deceive him into making mistakes and misjudgments had long been accepted among soldiers and statesmen. Then we dwelt on lesser versions of that same rationale when, as in poker or sports, one played out little deceptions to "fool the competition." Somewhat akin to these acts of deception are the little "white lies" that are employed on a regular basis to remove awkwardness from social situations, such as praising the hostess for her "marvelous" dinner.

Finally, I questioned Powers as to the impact of his continuing search for honesty on his career—would he recommend that others take the risk of getting low efficiency reports from seniors who preferred less rocking of the boat? "The risks," he said, "are not that great if you also have a reputation for competence in military duties. I was docked a couple of points but was not moved out of my outstanding ratings. All but one of the reports were maximums. In fact, the system of reports had become so scandalous that my last one declared 'This Report is *Not* Inflated.'

"The more important point is that a good OER is not the goal of the exercise. I learned at home in Illinois that a man's most precious possession is his reputation. I prefer my friends, colleagues, and particularly my family, to know and remember me for my honesty and straightforwardness.

"It was—and still is—great fun, not a burden, to build a reputation for competence and honesty. When I took over this National Guard company, I was required to sign for property that included clips of ammunition and 2643 loose rounds. I asked to see those loose rounds. Later, the warrant officer said that I was the first officer in 28 years to count those rounds. I told him that it was just an old habit of mine. I imagine the word got around about that.''

THE COMMANDER AS MORAL TRAINER

If we accept the Powers philosophy that both character and environment influence the commander's search for honesty in his command, then it follows that the commander will pursue certain tasks on a continuing basis. He becomes a constant trainer and teacher, especially in reaffirming the importance of character and interpreting moral principle into the concrete conditions of the command. He might even march the lieutenants to the theater and talk about Air Force headsets.

The commander also demonstrates his feelings toward honesty by taking actions that are visible to his people. It is not enough for a commander to be quietly honest. His beliefs must be overt, so that others can follow his example. He must announce policies that will help to establish a moral climate, such as clear guidelines on what constitutes right and wrong actions. His people will know him by the kinds of rewards and punishments he uses for proper and improper conduct. They will also judge him on his ability to keep competition and stress at levels that are short of generating cheating and lying among his subordinates.

The effective commander cannot be too far removed from the moral actions of his people, under the argument that "all of them are presumed honest until proven otherwise." In some units, this distant attitude will allow the least qualified to set the moral standard of the organization. Commanders need to create cross-checks on reports, and to establish periodic, impersonal audits and inspections, all of which occur automatically and do not imply that individuals or groups are being singled out for special investigation.

Even this short list of tasks presents some problems. How does one know if competition and stress are at the right level? Powers' battalion commander had one company cheating, but three not; is that too much, too little, or the right amount of pressure? We can also consider the commander who is told to support a program he does not believe in. Is he wrong in standing before the troops and praising it, even though he thinks he is misleading them?

What about the commander who is faced with a media reporter known for misrepresenting the facts in a manner hostile to the military; should the commander protect himself and the Army by his own evasions and misrepresentations? Concerning the commander in combat who is faced with low morale and probable defeat: is it wrong to tell the troops that they are invincible and near victory, if it means that they will make that one last supreme effort?

These questions, being typical moral dilemmas, have no fixed answers, nor even best answers. They have least-worst answers every time they are asked. They require pre-thinking. That is, they require reading, discussion, and reflection before being encountered. They require a settled philosophy concerning one's own moral make-up, and the moral capabilities of those with whom one is dealing.

For example, it is useful to have pre-thought the question of how humankind generally confronts moral dilemmas in their daily lives. A reading of Lawrence Kohlberg's *Essays on Moral Development* suggests that a commander who is making an appeal for more honest reporting may get different reactions from his troops, according to their different stages of moral development. He will have some soldiers at the most primitive stages of moral growth, where moral conduct is ruled by considerations of "What's in it for me?"—that is, by fear of punishment or hopes of immediate reward, or by some favorable exchange of favors.

On the other hand, he may have some subordinates who, because of their education and experience, are at the highest stages of moral development. They act to uphold values and principles, whether stated in law or a social contract like the Constitution, or as broad abstract principles, such as justice, equality, and respect for the dignity of the individual.

Between these extremes, most subordinates fall into the middle categories of moral development, where moral conduct is undertaken in order to maintain the expectations of the group, the family, or the nation. By acting morally, one conforms to the social order, demonstrates loyalty to authority, and "does one's duty."

Kohlberg's analysis indicates the need for a commander to present several levels of appeal, with strong emphasis on restating and supporting the core policies of the Army and the profession. This had not been done sufficiently by the commanders of the unit that perpetrated the massacre of Vietnam citizens at My Lai. Kohlberg interviewed all participants and found that the one enlisted man who clearly refused to fire his weapon tested out at the highest stages of moral development. He said he acted on principles that he had acquired in soldier basic training.

LEARNING ABOUT LYING

One does not explore lying and deception without finding that there are few valid authorities on the subject. Lieutenant Powers formulated his ideas and practices through his own experience and what others had told him. He did not read a single book, nor consult any analyst of military life or critic of American contemporary society who could lend sanction to his thoughts. He knew of regulations, laws, sermons, and essays that declared in absolutist fashion that deception and lying were evil and punishable. But he knew of no authority that defined these terms, addressed the extent of their practice, or described the processes through which humans cope with their effects.

In 1978, however, Sissela Bok undertook the publication of *Lying: Moral Choice in Public and Private Life,* a book that can provide guidance to future commanders in search of honesty in the military. In her work as Professor of Philosophy at Harvard University, she explored the writings of Aristotle, Augustine, Aquinas, Bacon, Kant and the modern philosophers. She applied their thinking to contemporary American society, both in those face-to-face relationships of everyday life and in the matters of grand statecraft. She explored military lying about U2 spyplanes, Cambodian incursions, and U.S. Army body-count symbols of victory.

Bok finds that lying is one of many forms of deception, which also includes evasion, exaggeration, disguise, silence, and inaction when these are designed to mislead. Lying requires the making of a statement in words, hand signals, or body gestures, with the intent of misleading or deceiving. Lying cannot be accidental. A person may become uninformed in many other ways, such as by a mirage. Some lies can be justified; most cannot.

Bok supports Powers' view that lying is prevalent throughout our society, and the military is no exception. "We lead our lives amidst all forms of duplicity," she says, "and from childhood we develop ways of coping with them. We all know what it is to lie, to be told lies, to be falsely suspected of having lied." (28 and 242)

This is rather strong medicine for senior officers who were schooled that "the truth, the whole truth, and nothing but the truth" is the Army standard for all, at all times, in all situations. For them, lying is a rare disease that must be cut out like a cancer; liars are to be discovered and purged. Powers and many of his cohorts see this reaching for the ideal world as quite noble. It becomes, however, a rationale for calling lying by some other name, and an excuse for marching lieutenants to the lecture hall rather than correcting the conditions that spawn deception.

Bok leaves no doubt that lying is evil, branded as a sin by theologians, and a violation of the fundamental moral principles that sustain civilized society. She believes that deceiving another is as evil as doing violence to another—both are assaults on human beings. But, being evil, lying has its appeal to all. She sees life as a continuing process of making moral choices between the apparent benefits of using deception and the liabilities of knowingly committing evil and having to pay a price. She describes how this price is so high that it is self-defeating of one's own goals, such as one's reputation for excellence or one's being able to live in a society of trust and mutual respect. "Trust and integrity are precious resources," she says, "easily squandered, hard to regain." (18 and 249)

What is necessary to justify a lie when one is about to undertake it? Bok doubts that one's own conscience is sufficient; it is best to get the opinion of other, hopefully disinterested, people. But what if it is in time of crisis, with no time to consult? She answers that the competent public servant will rarely come across situations that have not occurred before; his previous reading and study should tell him and his close associates the losses and gains he can expect from a public lie at that time. Bok presumes, however, that there should be much more extensive pre-thinking of crisis lying situations than we do now. Examples might be lying to prevent panic in a nuclear accident, lying to the enemy about the extent of damage he has caused in a bombing raid, or well-known situations where lying has done more harm than good, as in the Watergate fraud of the mid-seventies.

Bok is very sure of the culpability of institutions. "The social incentives to deceit are at present very powerful; the controls very weak. Many individuals feel caught up in practices they cannot change. It would be wishful thinking, therefore, to expect individuals to bring about major changes in the collective practices of deceit by themselves. Public and private institutions, with their enormous power to affect personal choice, must help alter the existing pressures and incentives." (244)

This gesture of sympathy for the individual caught up in modern institutions should not be interpreted as absolving them from any guilt for lying when under institutional pressures. Throughout the book, Professor Bok outlines why and how men and women should be constantly struggling to identify deceptive practices, to understand how and why they came about and, if they are leaders in institutions such as the Army, to take action to provide a new direction for all concerned.

After ten years of coping aggressively with dishonesty in the military, Captain Powers read Sissela Bok's *Lying*. He wished it had been available to him a decade earlier. "It is strong on the dilemmas one should consider in one's personal conduct, weaker on what organizational leaders need to do to reduce pressures for lying. If I had had this analysis in the beginning, I would have felt more comfortable in what I was doing, been a little braver about it, and known better how to answer my opposition."

The commander's search for honesty is more than a campaign to root out lying. The search starts with gaining an understanding of the nature of truth itself. Why is truth so valuable? How are men and organizations better off if truth prevails? Who says that truth is good? In addition to her book *Lying*, Sissela Bok provides answers in her 1984 *Secrets: On the Ethics of Concealment and Revelation*. More answers from philosophy are found in the first eight chapters of Mortimer J. Adler's *Six Great Ideas* and in the essay on truth in *The Great Ideas, A Syntopicon of The Great Books of the Western World*.

In recent years, many polemic tracts have been written about deception in public life, including journalist David Wise's *The Politics of Lying: Government Deception, Secrecy, and Power*. In *All The President's Men,* Carl Bernstein and Robert Woodward produced a bestseller about their investigation into the Watergate deception of President Nixon and his aides.

Lying by military people in wartime is not particularly well recorded in history and biography. The novelists of war, however, make it a central theme. Colonel Peter Stromberg, U.S.M.A. Professor of English, analyzed some fifty novels written about the Vietnam War while it was in progress. He found that the authors tested their soldiers in a variety of ways, especially in their capacities to exhibit honesty and integrity under the pressures of fighting an unpopular war against an implacable enemy in a distant tropical land. These fictitious commanders were required to stand up to four kinds of tests of character: honesty and integrity, skill in their work, compassion, and courage (with a willingness to sacrifice their lives). Many failed the test of honesty. Stromberg quotes Aeschylus, "In war, truth is the first casualty," and notes that, for the novelists of the Vietnam War, truth is a foremost target.

In Anton Myrer's *Once an Eagle,* Sam Damon passes the test of character, but Courtney Massengale fails. In Josiah Bunting's *The Lionheads,* General Lemming establishes the conditions that assure deception in the reporting of enemy body counts. Stromberg cites

Daniel Ford's *Incident at Muc Wa,* whose Captain Olivetti is portrayed as a professional officer who "is never conscious of any ideal of decency; he lacks integrity, courage, and compassion." (Stromberg, *A Long War's Writing,* 280.) Ultimately, he abandons his colleague and leaves him to die. Unfortunately, some of the best novels on Vietnam portray an Army that seems leaderless or led by corrupt men. The heroic commanders are yet to be discovered.

It is this last point that underscores the difficulty facing aspiring commanders who would prepare for their roles as moral arbiters. Except for the written materials prepared for service school classes in military ethics and professionalism, the literature is sparse and often negative. They must turn to the biographies of commanders who felt strongly about their moral responsibilities, particularly "Stonewall" Jackson and Robert E. Lee.

But their best school is a living one—the men and women around them who have the integrity of character to speak and act in a moral manner. Observing and listening to these moral arbiters reveals the roots of their beliefs, the sources of their values, and the rationales for their actions. Observing and listening also reveals the high value placed on their personal reputations—or, as soldiers have said for centuries, on their personal Honor.

Reading for Moral Arbiters

Mortimer J. Adler, *Six Great Ideas*.
Carl Bernstein and Robert Woodward, *All The President's Men*.
Sissela Bok, *Lying: Moral Choice in Public and Private Life*.
_____ , *Secrets: On the Ethics of Concealment and Revelation*.
Josiah Bunting, *The Lionheads*.
Daniel Ford, *Incident at Muc Wa*.
The Great Ideas: A Syntopicon of the Great Books of the Western World.
Lawrence Kohlberg, *The Philosophy of Moral Development: Moral Stages and the Idea of Justice*.
Anton Myrer, *Once an Eagle*.
Peter L. Stromberg, *A Long War's Writing: American Novels About the Fighting in Vietnam Written While Americans Fought*.
Malham H. Wakin, *War, Morality, and the Military Profession*.
Daniel Wise, *The Politics of Lying: Government Deception, Secrecy, and Power*.

CHAPTER SEVEN

The Commander's Concept of Duty

Duty, Honor, Country. These three hallowed words reverently dictate what you ought to be, what you can be, what you will be.

—General of the Army Douglas MacArthur
Thayer Award Address, May, 1962.

Shortly before the 1942 American invasion of North Africa, the Task Force commander, George S. Patton, wrote in his diary, "I hope that, whatever comes up, I shall be able to do my full duty. If I can do that, I have nothing more to ask. Fate will deliver what success I shall attain . . ." Three centuries earlier, the Frenchman Pierre Corneille advised in *El Cid,* "Do your duty, and leave the rest to heaven." Since the Age of Pericles, philosophers, playwrights, and generals have never doubted that duty was the central virtue of the professional military man. But this was not so in 1984, when two Washington study groups wrote 500-word statements of philosophy for Army systems that governed officer personnel management and professional development—never using the word Duty. Moreover, they did not mention Honor or Country. Instead, they wrote of commitment, selfless service, loyalty, and candor.

Was this a mere substitution of modern words for antique ones? Or was there a new message, a departure from a long tradition?

Subordinating "duty" to newer Army values seems to have evolved from the American experience in Vietnam. In 1980, the Chief of Staff of the Army, General Edward C. Meyer, spoke and wrote of Army values, emphasizing not duty but, rather, loyalty to institution, loyalty to unit, personal responsibility, and selfless service. A year later, General Donn A. Starry also omitted duty from his assessment of important military values; he expanded the Chief's list to include competence, commitment, candor, and courage. ("In Pursuit of an Ethic," *Army*, Sep., 1981, 11) The goal of this new language was to correct an evil that had come upon the Army—too many officers misbehaving and thinking it necessary to lie, cheat, and steal in order to get ahead.

The new language might well achieve this goal, for it underscored those values that would assure obedience to authority, be it to regulations, superior commanders, or institutions. The new words said little about mutual trust and obligations, about the professional growth of officers and NCO's, or about the importance of the strong individual in creativity, leadership, and command. The new language was the answer to a special problem in the Officer Corps. It was apparently necessary, because the old traditional concepts were too difficult to be taught and grasped by young people from contemporary American society.

The substantive difference between old and new was in the concept of self, the worth of the person, and the place of the individual in a shared human enterprise. While the new word "commitment," for example, implied giving over one's will to the cause (be it institution, ideal, or group), the old word "duty" implied that the individual should determine the nature and extent of his obligation, and then give the obedience and allegiance that reason dictated.

While the new "candor" called for truthfulness and frankness, it did so as an institutional requirement, for automatic conformance by the individual involved. The old word "honor" called for truthfulness and honesty to sustain, not only the institution, but the honor or reputation of the individual, whose most valuable asset was his good name for integrity and trustworthiness.

In the new language, loyalty to institution and unit became a requirement of conduct boldly demanded by superiors. In the old language, loyalty was subordinated to other values, a commodity to be earned by authority, and then offered as a duty by the individual because the object of his loyalty merited it.

Finally, the introduction of the new concept of "selfless service" seemed to advocate that the strong, self-centered personalities like Patton and MacArthur be avoided, as subversive to loyalty and good morals. In contrast, the old language did not necessarily probe into the motivations of the individual, but assumed that selfishness was ever-present and ineradicable, and that the self, while inviolable, was always in need of discipline, restraint, and temperance. The old philosophy wisely coped with "weakness of character," rather than foolishly trying to order evil into oblivion. In so doing, it nurtured the strong-willed and often self-centered personalities that breathed creativity, leadership, and discipline into Army life. The strength and meaning of these "old school" military values were delineated by Edgar F. Puryear in *Nineteen Stars,* a comparison of the ideals and careers of Marshall, MacArthur, Patton, and Eisenhower.

The future commander, looking for a settled philosophy about military conduct, might well ask whether he would prefer the old or new language in counseling subordinates. If he accuses a wayward lieutenant of not giving "selfless service" in the performance of his tasks, the junior officer may well assume that he has been insulted, ask for proof, and effectively terminate the conversation. If, on the other hand, the commander suggests that he and the lieutenant talk about the latter's conception of what his duty is and how it might be performed, no insult about motives need be inferred and some progress might be made. In the traditional language of duty, personalities could be left out, and an impartial discussion of a third entity entertained.

In the early chapters of these commentaries on military command, the lieutenant who thought his commanders too self-centered might have had a different view if he had been trained to think in terms of duty rather than of selfless service. If asked, "Does your commander carry out his duty in a worthy manner?", the lieutenant would have had to assess the nature of the senior's duty and whether he handled it well. The old language provided for distance between ranks, more suitable for cool assessment and dispassionate appraisal.

For the commander who would organize his thoughts and actions around the concept of duty, military memoirs and biographies offer a wide array of mentors. *The Eisenhower Diaries,* for example, reveal a five-star general's arguing within himself as to whether he should be a candidate for the Presidency. In August, 1951, as Supreme Allied Commander in Europe, Ike wrote that Republican leaders had come to persuade him to run for the job.

They recognize that I have an important duty in this post. They believe that I have (rather, will have) a more important duty, to accept the

Republican nomination. . . . I've told them, as I tell all, that I'll certainly always try to do my duty to the country, when I know what that duty is. As of now I have a duty; I cannot yet even describe the circumstances that would be conclusive in convincing me that my duty had changed to that of assuming a role in a political field. (198–199)

Two months later he wrote:

I entered upon this post only from a sense of duty—I certainly had to sacrifice much in the way of personal convenience, advantage, and congenial constructive work when I left New York [as president of Columbia University]. I will never leave this post for any other governmental task except in response to a clear call to duty. I will not be a participant in any movement that attempts to secure for me a nomination because I believe that the presidency is something that should never be sought, just as I believe, of course, that it could never be refused. . . . I would consider the nomination of which they speak, if accomplished without any direct or indirect assistance or connivance on my part, to place upon me a transcendant duty. . . . As of now I see nothing to do but keep my mouth shut. (204)

In the early part of 1952, Ike agreed that a nomination by the Republican Party constituted a valid call to a more important duty. His diary entries reflected three characteristics of the person who has guided his life toward a star of duty. First was his ingrained desire to do the right thing—to obey the law, to meet obligations, to await a mandate from others. Second was his determination to uphold principles that he had adopted for himself—to serve his country, and keep out of politics when in uniform. Third was his awareness that one has many duties, which may often be in conflict. One must make choices about which duty is transcendant at any given time.

THE NATURE OF SOLDIERLY DUTY

While few men have the burden of deciding if it is their duty to become President of the United States, all Army officers face challenges to their sense of duty every day. At the working levels of the Army, the questions are quite mundane: Is it my duty to see that this Hispanic soldier learns to read English? Is it my duty to volunteer for a combat assignment? Is it my duty to stay in the company overnight, to see that we are ready for tomorrow's inspection? Is it my duty to stand in for a fellow officer when he is unaccountably absent?

Each of these questions arises from the demands of duty which, for military people, has at least eight faces or varieties that must be honored. Chart 5 is a quick summary of the scope of the soldier's duty.

The first four varieties of duty stem from the officer's commissioning warrant, the oath of office, and acceptance into a profession with a special responsibility. The remainder are the responsibilities of any human being, although they take on greater meaning because of the public nature of the officer's work; it is the duty of military leaders to set an example. They do not have the freedom of civilians to ignore these charges. Only the eighth duty is self-selected, since it is for those officers whose God and religion are a fundamental allegiance in their lives; they adhere to Ecclesiastes XII, 13: "Fear God and keep His Commandments, for this is the whole duty of man."

The Chart's listing of the varieties of a soldier's duty includes both professional and personal obligations. This is considerably at variance with those who would speak only of professional matters, leaving the officer's private life to "his own time." Human beings are not divided into such neat categories; they each have one mind and one character,

Chart 5

The Varieties of the Soldier's Duty

1. The duty to obey orders and carry out assigned missions.

2. The duty to care for subordinates and build strong units.

3. The duty to defend the nation, its people, and its values.

4. The duty to uphold the profession and its ethical code.

5. The duty of self-development to one's highest potential.

6. The duty of fidelity to family, friends, and colleagues.

7. The duty to uphold the moral principles of civilized man.

8. The duty to one's God and religion.

and weigh their decisions and actions as a total person. To some degree, officers are "on duty" twenty-four hours a day, and are never entirely professional or entirely personal. Rather, military people must consider all facets of their duty concurrently, as though holding a cut diamond to the light and looking at each of its facets as an expression of the whole.

It is normal for these varieties of duty to be in conflict with each other. For those concerned with doing their duty, life is a succession of choices that relegate the myriad of obligations and duties to their proper time, place, and importance. Often, choices result in compromise. The combat commander, for example, may estimate that his mission will produce excessive casualties if conducted according to the operations order; he may seek to resolve this conflict between duty to mission and duty to the troops by requesting an alternative timing or route of advance, which can complete the same mission with fewer casualties.

In 1951, General Douglas MacArthur, who spoke more about duty than any other soldier, felt that he had to choose between obeying the orders of President Truman and defending the nation's interest by waging war more aggressively against the Chinese. He found no compromise, persisted in his dogma, and was relieved of his command. In succeeding years, he argued that the soldier's first duty is to the Constitution rather than to the men temporarily in power. The same "higher duty" argument was used by officers dissenting from service in Vietnam. This pressing problem of disobedience in the sixties is analyzed in Michael Walzer's *Obligations: Essays on Disobedience, War, and Citizenship*.

When claims of duty are used to govern decisions and actions, the soldier becomes aware of the limits and terminal points of one's duty. Like MacArthur, some take their dissent to the point of giving up their commissions and leaving active service. More customary is the decision to limit one's dissent at the point where argument over policy has ceased, a decision is made, and unity is needed to carry out the policy.

Another important limitation of the duty to obey is the code clearly stating that an officer should not obey an unlawful order. Whether one can prove that the order is or could be illegal may be very difficult. Unless the issue involves murder or some other equally heinous crime, the refusal to obey is usually worked out between contending parties before a court must decide whether or not an order is legal. The important effect of the code is to deter commanders from issuing illegal orders, rather than to safeguard those who disobey.

The most significant limitations of the soldier's duty are those that must be self-imposed by the soldier himself. Duty to the family cannot be met fully if one's duty to subordinates and units is to be fulfilled effectively. Each professional finds his own individual compromises between these claims. Soldiers often feel that they must limit the duty owed to friends and colleagues, particularly when such friends misuse a friendship to ask for favors bordering on unethical, illegal, or immoral activities. In these cases, one "discharges" the duty with which one has previously "charged" oneself.

How do commanders judge between the demands of conflicting duties? Some do it well; others, poorly. Many barely recognize that the problem exists, and are continually surprised to find their decisions ill-advised and not carried out. These commanders are usually out of touch with their people, although they may talk with them every day. They appear to be without strong moral convictions, are easily swayed by the last person they talked to, and are pliable in the hands of their seniors. They react to fear for their popularity and to opportunities for immediate gain. They are at the beginning stages of moral growth, as suggested in Lawrence Kohlberg's *Essays on Moral Development: Moral Stages and the Idea of Justice*.

The majority of American military commanders fit into Kohlberg's intermediate moral stages much better; they adhere to the set of moral principles brought to the Army from their early years of family and religious training. They make their judgments heavily in favor of the values of the institutions that they serve—the Army, the nation, and the units to which they are assigned. Obedience, patriotism, loyalty to the mission and the unit: these are the duties weighing the most in their judgments. For some, however, the results are often crude and misguided—decisions that reflect the jargon of the trade, such as "I go by the book," and "We accept only zero defects." This narrow orientation produces old responses to new challenges, a failure to serve the real needs of the Army, and a cookie-cutter standard reaction to problems rather than the creative solutions that are needed.

In contrast, the best commanders tend to put strong emphasis on their duty "to uphold the moral principles of civilized man," using it to make judgments about the many varieties of duty they face. They are aware of the moral principles that have been passed down through the ages. They believe in the efficacy of these principles in sustaining worthy communities and institutions, and act in accordance with them. Through this knowing, believing, and doing, they are identified with

Chart 6

Moral Principles of Primary Concern to Military Commanders

Beneficence towards others, as reflected in:

- not inflicting evil or harm on others, killing, stealing, bearing false witness
- preventing evil or harm from coming to others
- removing evil from institutions and society

Fidelity towards others, as reflected in:

- truth-telling
- promise-keeping
- reparation for previous wrongful acts
- gratitude for previous beneficial acts, such as giving and preserving life.

Justice towards others, as in assuring a proper distribution of rights, benefits, and injuries

---

Kohlberg's higher levels of moral development. Chart 6 presents a listing of these moral principles, derived from Dr. Arthur Dyke's *On Human Care,* a textbook that has been used in courses in military ethics at the U.S. Army War college.

It is evident from this Chart that moral principles have to do with one's actions towards other human beings, whether beneficence, fidelity, or justice. These "rules" govern the actions of men and women in dealing with each other as human beings—as part of a society of mankind. Commanders who adhere to these principles see themselves as leaders of a community over which they have the authority to act. Here there is little room for the ego-gratification or self-aggrandizement that afflicts commanders who see themselves standing in splendid isolation above the masses.

PERFORMING DUTY FOR REASONS OF JUSTICE

Cadets who came into the Army in the fifties often listened to the following story told by a chaplain. There was once a traveler on the road to Jericho who had been beaten, robbed, and left to die. Two noble citizens passed him by as if they did not see him. Then a Samaritan, a stranger in the land, came by. He stopped, dressed the traveler's wounds, and took him to an inn to recover. The chaplain finished his story of The Good Samaritan with an observation. Had there been a soldier stationed on the road to Jericho, whose duty was to protect travelers from evil, the victim would have never been assaulted, the passersby would not have been mocked for their hypocrisy, and the Samaritan would not have been tested for his compassion, nor delayed on his journey. The cadets left the sermon convinced that the life of the soldier was worthy, made so by his duty to safeguard the citizenry of a nation.

But the chaplain did not go quite far enough. How do we know that the soldier posted on the road to protect travelers would actually carry out his duty? Might not the soldier ignore the robbers if they were better armed than he? Was there any guarantee that he would not join them in their plundering? Or suppose that a group of citizens came to the soldier one day, saying, "The government is corrupt and exploits the people. You have more important things to do than to protect this road. It is your higher duty to join us and employ your weapons to pull down this evil government." What, if anything, will cause the soldier to resist such temptations?

Obedience and discipline, if properly understood and enforced, can keep the soldier aware of his duty in ordinary times. But the commitment to duty that is necessary under exceptional circumstances must be reinforced by values that tie the soldier to the service of other human beings. Chief among these soldierly values is a sense of justice.

In 1939, General Sir Archibald Wavell foretold an audience at Cambridge University:

> In a future war . . . discipline should be a different matter from the old traditional military discipline. It has changed greatly since I joined, and is changing still. But, whatever the system, it is the general's business to see *justice* done. The soldier does not mind a severe code, provided it is administered fairly and reasonably.

> —Wavell, *Soldiers and Soldiering,* 28.

Like Wavell, most soldiers equate justice with fairness, the fairness of one's dealing with other human beings. This is *distributive justice,* wherein equal individuals merit equal shares of social goods and evils. The commander, however, is also concerned with two other kinds of justice. One is *contributive justice,* wherein each person must contribute equally to the sustaining of institutions that benefit him, such as his government or his infantry company. The other is the justice of *human rights,* wherein every human being is entitled to certain inalienable rights, such as freedom from murder, bondage, theft, and libel. The legal justice that must be provided by the military commander is largely in the justice of human rights. The inquiring commander finds further analysis of these three concepts of justice in Mortimer J. Adler's *Six Great Ideas* and John Rawls' *A Theory of Justice.*

Aristotle argued that man is governed by love in his dealings with his family and very close friends. But beyond this tight inner circle, in the absence of love, man must be governed by justice, as promulgated by a tribal, institutional, or state government. Through justice, each receives according to what is due him; the determination of "just dues" within the organization may be based on law, merit, or fairness.

Not everyone agrees with Aristotle that rewards should be based on merit, and that those who are unequal should receive unequal shares of the rewards or punishments. By the test of fairness, however, each man is entitled to equal treatment with others, and if some are given disproportionate shares of material wealth, power, or freedom, this inequality must be justified on the grounds that it benefits everyone in the organization. Hence, a commander acquires more power or physical comforts than that of his subordinates only if the least advantaged man in the organization benefits from this unequal distribution.

Justice is said to be the first virtue of social institutions, just as truth is of systems of thought. Wise commanders of military "social institutions" know that justice demands fairness as well as lawfulness, and that they must insist that every soldier willingly contributes his fair share of the work to the cooperative enterprise of the unit. But they also know that every obligation of the soldier has an earned right attached to it. These are obverse sides of the same coin, and for every sacrifice of freedom of movement and speech that a soldier makes, he can rightfully expect commanders to use restraint in curbing the soldier's fundamental liberty.

Understanding the citizen-soldier's right to basic freedoms has been most difficult for American commanders of the past two generations. All citizens are guaranteed freedom of thought and conscience; freedom of speech, assembly, and personal movement; freedom to participate equally in political affairs; and a variety of basic civil liberties. But if men are to act as one in a fighting organization, obedient to civilian and military command and able to live together in close quarters without harming each other, they are required to forego some of their inherent freedoms.

The wise commander, however, knows that subordinates needs sufficient freedom to develop their capacities and to avoid losing their identities as men and women. He also knows that the protection of this freedom is the fundamental reason for America's wars with Naziism, Japanese militarism, and communism in Korea and Vietnam. Such commanders recognize that to curb the soldier's liberty beyond what is demanded by real military necessity is to perpetrate injustice and unfairness among the troops. Commanders who substitute personal whim and craving for power for military necessity in determining the liberties of their subordinates suffer the consequences—a decline in unit morale and effectiveness.

The respect for an individual's dignity that causes the commander to give a subordinate extensive freedom also causes him to resist the temptation to manipulate the soldier into shameful or wrongful deeds. To lead, we are told, is to cause others to do willingly what one wants them to do. Commanders who are unused to power find security in manipulating men through fear, intimidation, or false promises, thereby manipulating them into "willing" performances. Such commanders may misunderstand the writings of psychologists (such as B.F. Skinner in *About Behaviorism)* that describe how researchers can manipulate their human objects into certain behavioral patterns. Commanders who seek "the good" of their troops, however, are more prone to look upon manipulation of subordinates as the philosophers do; that is, as unethical and an assault upon the integrity and worth of other human beings. The trick is, of course, to distinguish between proper demands on subordinates and the unethical manipulation of them. A concern for justice and fairness makes this discernment easier.

In 1804 Napoleon wrote, "There is no strength without justice." Today, the strength of the American Army is sapped by the injustice that arises from neglect or miscalculation—errors in pay, promotion, rotation of assignments. Correcting injustice here is a matter of sensing the problem and pursuing it with skill through the labrynthine system. Much of the injustice is also due to bad policy decisions, ordering conflicting requirements without providing the time and resources for accomplishment, until men are senselessly overworked, deprived of deserved liberty, or wrongly rated on their performances. These injustices call for a quick policy change; only commanders who are aware of the real price paid for injustice can avoid or correct such wrongdoing.

When neither miscalculations nor erroneous policy are to blame, we look for the villain who forces injustice on subordinates for a variety of ego-gratifying reasons—the desire to see lackeys jump, the urge to keep other races or ethnic groups in servility, the hankering to savor anachronistic rules long after they have become unjust, only because they are symbolic of a more comfortable past. Nevertheless, the villains argue that they are only acting in the organization's best interest, in which they crave a more generous share of power and privilege. It is most difficult to avoid or correct injustice coming from this source. But the process of eradicating these miscreants goes on. In the words of the Naval Board investigating the mutiny on H.M.S. Bounty: "If justice be not in the mind of the Captain, it be not aboard."

THE DUTY OF SELF-DEVELOPMENT

On 7 July, 1970, Captain John Alexander Hottell was strapped in a helicopter that was caught up in a tropical storm and slammed into a hillside in a remote mountain area of Vietnam. Shortly before, while commanding a company of the 1st Cavalry Division, he had written a sealed letter to his wife, Linda, which began:

> I am writing my own obituary . . . [because] I am quite simply the last authority on my own death.

> I loved the Army: it reared me, it nurtured me, and it gave me the most satisfying years of my life. Thanks to it I have lived an entire lifetime in 26 years. It is only fitting that I should die in its service. We all have but one death to spend, and insofar as it can have any meaning it finds it in the service of comrades in arms.

> And yet, I deny that I died FOR anything—not my Country, not my Army, not my fellow man, none of these things. I LIVED for these things, and the manner in which I chose to do it involved the very real chance that I would die in the execution of my duties. I knew this and accepted it, but my love for West Point and the Army was great enough— and the promise that I would someday be able to serve all the ideals that meant anything to me through it was great enough—for me to

accept this possibility as a part of a price which must be paid for all things of great value. If there is nothing worth dying for—in this sense—there is nothing worth living for.

The Army let me live in Japan, Germany, and England, with experiences in all of these places that others only dream about. . . . I have climbed Mount Fuji, visited the ruins of Athens, Ephesus, and Rome . . . and earned a master's degree in a foreign university. I have known what it is like to be married to a fine and wonderful woman and to love her beyond bearing with the sure knowledge that she loves me; I have commanded a company and been a father, priest, income-tax advisor, confessor, and judge for 200 men at a time; I have played college football and rugby, won the British national Diving Championship two years in a row, boxed for Oxford against Cambridge only to be knocked out in the first round. . . . I have been an exchange student at the German Military Academy, and gone to the German Jumpmaster school. I have made thirty parachute jumps from everythiing from a balloon in England to a jet at Fort Bragg. I have written an article for *Army* magazine, and I have studied philosophy.

I have experienced all these things because I was in the Army and because I was an Army brat. The Army is my life, it is such a part of what I was that what happened is the logical outcome of the life I lived. I never knew what it was to fail, I never knew what it is to be too old or too tired to do anything. I lived a full life in the Army, and it has exacted the price. It is only just.

Just, yes, in the personal philosophy of Alex Hottell. When the obituary eventually appeared in the press, it was admired for its expression of gratitude for being permitted such opportunities to learn, and for the sense of justice in the tradeoff of a life well lived for a cause well merited. Some felt that it expressed well the sixties' absorption with self. They could not, however, attribute selfishness to his willingness to give up life in fulfillment of that self. This was not "selfless service," but "self-in-service."

Hottell had been an excellent commander at the company level. He had earned two Silver Stars, and some said that he had been "rescued from himself" by sudden orders to division headquarters. If there was an explanation for his nascent capacity to command, it lay in his incessant desire to learn—to read, to experience new phenomena, to question, to experiment. The duty of "self-development to one's highest potential," although stated crudely in Chart 5, symbolized Hottell's personal philosophy of daily living. His enforcement of it, however, was more in keeping with Theodore Ropp's reason for reading military history—"because it is fun."

Learning about the role of a military commander must start early enough for a person to acquire a vision of himself in command roles at several levels higher than then held. Hottell saw a model in Major General George W. Casey, who died in the crash with him: "He is imaginative, aggressive, charming, and has a more complete grasp of the complex missions that confront the American division commander than I would have thought possible. It will be almost a religious experience for me to serve with the Cav when this man commands it." Such visions of what a commander should be came not only from Hottell's reading of military biography, but also from his study of philosophy and literature; he had read not only Plato's description of the "man of virtue," but also the modern expressions of the same theme, such as John F. Kennedy's *Profiles in Courage.*

The duty of self-development for commanders seems to call for three study objectives. First is the acquisition of knowledge and skills associated with the several roles of the commander—leader, manager, tactician, warrior, strategist, and moral standard bearer. Second is the acquisition of knowledge, insights, and values associated with the virtuous human being, perhaps best stated in Plato's ideal of the man of wisdom, courage, temperance, and justice. Third is the acquisition of insights gained from thought about oneself and the personal style that is suited best to a commander's role in the twentieth century Army environment.

For pursuing the first of these objectives, the early chapters of these commentaries cite books useful in developing one's vision as a tactician, warrior, professional, leader, and manager. In addition to these, future commanders will want to read the recent mind-expanding popular works that widen the vision of the management expert, such as John Naisbit's *Megatrends: Ten New Directions Transforming Our Lives;* Scott and Hart's *Organizational America;* and Peters and Austin's *A Passion for Excellence.* In the field of leadership, some of the best works deal with political leaders, such as James McGregor Burns' *Leadership,* Field Marshal Bernard Montgomery's *The Path to Leadership,* and Dankwart A. Rustow's *Philosophers and Kings: Studies in Leadership.*

In order to develop a more complete understanding of Plato's "virtuous leader," commanders must turn to some of the classic and popular works of philosophy, religion, literature, and history. Mortimer J. Adler has made some of the best ideas on justice and duty readable in *Six Great Ideas* and *Aristotle For Everybody: Difficult Thought Made Easy*. For a remarkably penetrating analysis of the relationship between justice and man's ability to cope with modern society, there is Alasdair MacIntyre's *After Virtue: A Study of Moral Theory*. Recent popular works exploring the problem of self and its place in the philosophy of public leaders include Erich Fromm's *Man For Himself* and C.S. Lewis' *The Abolition of Man*. The problems of holding power, with its dangers of corruption by hubris and arrogance, are explored by Thucydides, Macchiavelli, Hobbes, Lord Acton, and the fathers of the American Revolution; David Kipnis attempts to pull these ideas into some understandable pattern in *The Powerholders*.

The most dramatic portrayals of military commanders who have lost track of their sense of justice and the reality of their duty are Cecil Woodham-Smith's *The Reason Why* and C.S. Forester's *The General*. In contrast, Forester's story of *Rifelman Dodd* portrays the trials of a private soldier trapped behind the lines in the Napoleonic Wars, and driven to heroic feats by his sense of duty.

The third objective for commander self-development is the most difficult. It requires the creative generation of insights about one's strengths and limitations, and the finding of a personal style of command that best matches one's personality with the demands of the job in the reality of modern times. Not all are to be Pattons, some might be Marshalls, none should be Custers. All should assume that command may be thrust on them in time of need; many should step aside, so that commanders of greater potential can get experience.

Assessing self, present and potential, requires the intake of ideas generated by first-hand experience, and by the vicarious experience of reading. But the gestation and analysis of this intake often requires writing. "I do not know what I think until I try to write it down," is the byword of men and women intent on probing into their beliefs and competencies. Nearly all the inquiring soldiers discussed in these commentaries in military command have been writers as well as readers; Clark and Hottell started personal journals before entering cadet training. Writing articles for professional periodicals seems to be standard fare for future commanders. Senior commanders finish their careers by publishing books. All reflect a lifetime of note-taking, research, experimental speeches, records of conversations, summaries of reading, diaries of current thoughts, file-drawers of trial paragraphs and short talks, and "commonplace books" of poems, quotations, and speeches. These are the tools of inquiring minds, who look upon books as fuel for the mind, as gasoline is to the internal combustion engine. Reading is a means, not an end; expressing one's thoughts is the end. Xenophon's *Anabasis,* Caesar's *Commentaries,* and Napoleon's *Maxims* are summary statements of thinker-doers who were writers. Modern military men and women who are writers customarily have a reference shelf of best books alongside their dictionaries; there is a distillation of these military writers' reference collections at the end of this chapter.

Alexander Pope wrote that the proper study of mankind is man. It then follows that the proper study of military command is military commanders. The biographies and memoirs described in Chapter 2 provide such a study. There is, however, a special way of thinking about other men's battle experiences if one asks "How am I suited for this and what is my best style?" Questions form in the reader's mind, and he asks them of the biographies and memoirs, which then become a goldmine of research, as well as a task undertaken "because it is fun." Each researcher has his own questions, based on his own particular inventory of hopes, fears, biases, and ambitions. In the final analysis, it remains true that there is no stereotypical commander around whom all must model their lives. Command is so unique a blend of personality and task that no two commanders should mirror each other. To the extent that they share common qualities, it is in their common search for courage, truth, duty, and justice.

Reading About the
Concept of Duty

Mortimer J. Adler, *Aristotle for Everybody: Difficult Thought Made Easy.*
_____, *Six Great Ideas.*
James McGregor Burns, *Leadership.*
Arthur J. Dyke, *On Human Care: An Introduction to Ethics.*
Dwight D. Eisenhower, *The Eisenhower Diaries.*
C.S. Forester, *The General.*
_____, *Rifleman Dodd,* also printed under the title *Death to the French.*
Erich Fromm, *Man for Himself.*
Colonel G.F.R. Henderson, *Stonewall Jackson and the American Civil War.*
John F. Kennedy, *Profiles in Courage.*
David Kipnis, *The Powerholders.*
Lawrence Kohlberg, *The Philosophy of Moral Development: Moral Stages and the Idea of Justice.*
C.S. Lewis, *The Abolition of Man.*
Alasdair MacIntyre, *After Virtue: A Study in Moral Theory.*
Bernard L. Montgomery, *The Path to Leadership.*
John Naisbitt, *Megatrends: Ten New Directions Transforming Our Lives.*
Thomas J. Peters and Nancy Austin, *A Passion for Excellence.*
Edgar F. Puryear, Jr., *Nineteen Stars.*
John Rawls, *A Theory of Justice.*
Dankwart A. Rustow, *Philosophers and Kings: Studies in Leadership.*
William Scott and David Hart, *Organizational America.*
B.F. Skinner, *About Behaviorism.*
Michael Walzer, *Obligation: Essays on Disobedience, War and Citizenship.*
FM Archibald P. Wavell, *Soldiers and Soldiering.*
Cecil Woodham-Smith, *The Reason Why.*

A Reference Shelf for the Military Writer

Jacques Barzun and Henry F. Graff, *The Modern Researcher*.
Lawrence P. Crocker, *The Army Officer's Guide*. 42d Ed.
The Great Ideas: A Syntopicon of Great Books of the Western World. 2 Vols.
Thomas E. Griess (ed), *The West Point Military History Series*.
Robert D. Heinl, Jr., *Dictionary of Military and Naval Quotations*.
John E. Jessup, Jr. and Robert W. Coakley (eds), *A Guide to the Study and Use of Military History*.
Anthony Kellett, *Combat Motivation: The Behavior of Soldiers in Battle*.
Jon Stallworthy (ed), *The Oxford Book of War Poetry*.
U.S. Army Center of Military History, *Style Guide*.

CHAPTER EIGHT

The Commander as Strategist

A professional soldier is rarely a professional strategist.

—B.H. Liddell Hart, *Thoughts on War*, 231.

When he made this observation in 1933, Basil Liddell Hart attributed the strategic failures of professional soldiers to their one-sided military training. He said this was concentrated almost exclusively on "the mechanisms of tactics and military organization." He might have included professional politicans on the same grounds. Although they may be devoid of all training in the art and science of war, they can be even more important than soldiers in the development of a nation's strategy.

Liddell Hart did not foresee the success of some of the Allied commander-strategists in World War II, such as George Marshall and Douglas MacArthur. Recent writing on the American performance in the Vietnam War, however, seems to reconfirm the lack of insightful judgment of those who shape American military policy and strategy.

Is it true that professional soldiers are rarely competent strategists? What does this say to the military reader who strives for excellence in command assignments? Must soldiers expect to participate in the formulation and execution of strategy—and fail because they have been so busy in their tactical and managerial roles?

What is strategy, and how is it conceived? Do only senior generals participate in its design? Who carries out strategic plans? What constitutes success or failure in strategic missions? What learning is essential to the soldier with strategic responsibilities? What knowledge, skills, and insights are necessary? What kind of reading supports this?

These are the questions we hope to explore in this final chapter of *The Challenge of Command*.

THE DIMENSIONS OF STRATEGY

Scholars and statesmen have been redefining strategy ever since the ancient Greeks identified *strategos* as "the art of the general." In the twentieth century, the generals and admirals of the American Joint Chiefs of Staff defined their work in this way: "Military strategy is the art and science of employing the armed forces of a nation to secure the objectives of national policy by the application of force, or threat of force." (USJCS, *Dictionary of Military and Associated Terms*, 217) This definition linked military force with policy, as Liddell Hart did in 1954. He called strategy "the art of distributing and applying military means to fulfill the ends of policy." (Liddell Hart, *Strategy*, 321) At a level above military commanders, Liddell Hart expected statesmen to establish the national policy and its political objectives. To achieve these "ends of policy," they should outline a *grand strategy* designed to coordinate and direct all the resources of a nation or band of nations— economic, diplomatic, psychological, technological, as well as military—towards attainment of the political goals.

The division of labor between political and military leadership was unnecessary when Alexander, Frederick, and Napoleon combined both roles. But in the nineteenth century, political leadership in the Western world became more exclusively civilian, and the military more professional. This required new definitions of responsibility—new formulations that have not always worked well. Historians write of the excessive influence of Junker militarists on the political goals of Imperial Germany, and of the Allied pursuit of "unconditional surrender" at the expense of more judicious objectives in World War II. Contemporary critics of American conduct of the war in Vietnam suggest that the political leadership failed to establish clear and realistic political objectives, and that their military advisors failed to develop a strategy that promised decisive results. As a result, the nation's strategy won the battles but lost the war.

In his 1982 *On Strategy: A Critical Analysis of the Vietnam War,* Colonel Harry G. Summers, Jr. suggested that the political and military strategists might have fared better had they adhered more closely to the theorists of strategy, particularly Karl von Clausewitz. Summers contrasted timeless theory with American practice in selecting war goals, employing the principles of war, and allocating resources to the war. He concluded that "we failed to properly employ our armed forces so as to secure US national objectives in Vietnam. Our strategy failed the ultimate test . . . ." (25)

The theorists of strategy have not been sufficiently clear in their guidance to permit us to say, "Had the generals studied Clausewitz, Jomini, Liddell Hart, and Fuller, they would not have made these mistakes." Clausewitz' *On War* is like the Bible in that one can find support for many courses of action. It is a useful tool for analyzing wars and campaigns, but not a template for selecting the "approved solution" to any given strategic problem. Clausewitz wrote that theory "is meant to educate the mind of the future commander or, more accurately, to guide him in his self-education; not to accompany him to the battlefield." (*On War,* 141)

Furthermore, each theorist has been in search of the ideal single strategy to fit all wars, causing each to advocate an approach that could not be reconciled with the others. In most cases, Clausewitz sought the winning of war through the destruction of the enemy forces in climactic battle. Jomini sought victory by threatening the opponents' lines of communication. J.F.C. Fuller focused on military action that would secure an ideal and lasting peace after the war.

Liddell Hart's *Strategy* was subtitled "The Indirect Approach" because strategic ends were to be achieved by causing the enemy to fight on terms least favorable to him. Liddell Hart wrote that strategy's purpose "is to diminish the possibilities of resistance." Perfection in strategy, he said, would "produce a decision without any serious fighting. . . . Dislocation is the aim of strategy, its sequel may be either the enemy's dissolution or its easier disruption in battle." (323–325) The full development of this line of strategic analysis is the subject of Brian Bond's *Liddell Hart: A Study of His Military Thought.*

REQUIREMENTS FOR THE MASTER STRATEGIST

Knowing and being able to apply the theoretical bases for strategic decision-making is required of any soldier who expects to be adept at "the art of the general." What other knowledge and skills are necessary if one is to be among Liddell Hart's "rare" professional soldiers who are also professional strategists?

I asked this question of Colonel Paul L. Miles, who had a decade of experience teaching military history, including courses entitled "War and Its Philosophers" and "Grand Strategy in the Twentieth Century." In his first tour in Vietnam, his Army engineer company built its share of the logistics and naval base at Camranh Bay. In his second tour, he was a member of the Joint Military Commission of North and South Vietnamese, Viet Cong, and American officers, whose purpose was to implement provisions of the 1973 Paris peace treaty. He had studied history as a Rhodes Scholar at Oxford, and had served as Aide to Army Chief of Staff General William C. Westmoreland.

Colonel Miles listed several tasks or duties that military strategists must perform if they are to be adept at their calling:

1.   Understand and support political *goals,* to insure effective coordination of policy and strategy.

2.   Select military *objectives* that will lead logically to the achievement of political aims.

3.   Allocate military *resources* and establish correct priorities.

4.   Conduct war in a way that sustains *support* of the home front.

5.   Maintain a proportional *balance* between the application of violence and the value of the political goals.

These duties are not easily carried out. Miles said that the first task—understanding political aims in the war and a common outlook on the correlation of policy and military operations were necessary in meeting the "supreme test" of the statesman and military strategist: "to establish . . . the kind of war on which they are embarking, neither mistaking it for, nor trying to turn it into, something that is alien to its nature." (Clausewitz, *On War,* 88) Moreover, it was significant that even Jomini, who frequently overlooked the political factors in his commentary on strategy, had written: "The first care of [the] commander should be to agree with the head of the state upon the character of the war. . . ." (Jomini, *The Art of War,* 66)

Miles emphasized the requirement for effective communication be-

tween the heads of government and their military advisors and commanders. "If military leaders are to develop appropriate strategies
across the spectrum of war, from total to limited to low-intensity
conflict, they must be assertive in fostering open and frank discussions
of strategic alternatives with their civilian leaders. At the same time,
this process should be reciprocal—as Clausewitz observed, policy
should not be a 'tyrant.' The relationship between Lincoln and Grant
during the last phase of the Civil War was a classic example of the
kind of dialogue between a Commander in Chief and a Commanding
General that produced a decisive strategy. In this case, the result was
both the defeat of the Confederate armies *and* the early restoration of
the Union.

"President Roosevelt's military advisors—Leahy, Marshall, King,
Arnold—realized early in World War II that the immediate political
goal and overriding military objective was not only the unconditional
surrender of the Axis powers, but at 'the earliest possible date.' This
goal then became a kind of litmus test against which the President and
his advisors could measure alternative strategies. This common outlook
contrasted with the unfortunate controversy between President Truman
and General MacArthur during the Korean War. This episode, which
threatened a crisis in civil-military relations, stemmed largely from
MacArthur's public advocacy of an alternative strategy that was incon-
sistent with the political aims of the American government."

Miles again quoted Clausewitz (119): "'Everything in war is very
simple, but the simplest thing is difficult.' The selection of decisive
military objectives may appear easy in theory; in practice it has been
another matter. Strategists must not only identify the long-term objective that will lead to a successful termination of the war, but must also
select intermediate objectives for campaigns of subordinate commanders. This task becomes especially complex in coalition warfare, where
differences in historical tradition and theoretical approaches foster
heated debates between allies over campaign plans and priorities. For
example, during World War II, American and British strategists agreed
that the eventual defeat of the German army on the continent was a
prerequisite for unconditional surrender of the Third Reich. But agreement on the long-term objective did not preclude almost two years of
debate and compromise over intermediate objectives in North Africa
and the Mediterranean. In addition, there was the related issue of the
contribution of strategic air power to the final defeat of Germany.

"In the Mediterranean, the intermediate objectives for Eisenhower, the British commander, Alexander, and other Allied leaders were made frequently on the basis of feasibility rather than orthodox strategic doctrine. I have always thought that Michael Howard's *The Mediterranean Strategy in the Second World War* contained an especially succinct statement of this reality: 'The development of Allied strategy was a piecemeal affair, in which military leaders had often simply to do what they could, where they could, with the forces which they had to hand.'"

Recognizing that logistics so often dictates strategy, Miles encourages his students to read chapters of *Command Decisions,* edited by Kent Greenfield. In these studies of World War II, one finds the economic and logistical rationale for the strategy of "Germany First," and the influence of American manpower shortages on the Army's "90-Division Gamble."

Miles continued: "The crucial—but frequently neglected—factor for the American strategist is the role of the home front. American military leaders—servants of a democracy—are obligated to devise strategies that command public support and sustain national will. To underline this concept for students, I try to add the term 'incremental dividends' to their strategic lexicons. The most succinct historical commentary on the essence of the concept is found in Liddell Hart's chapter on the American Civil War in *Strategy*. According to Hart, it was the 'dividends' of Sherman's capture of Chattanooga, the occupation of Atlanta, and the March to the Sea that sustained the North's support for the theoretically more decisive, but also more costly, campaigns that Grant waged against Lee's army in Virginia. Hart's conclusion is one that our strategists should take to heart: 'The strategist who is the servant of a democratic government has less rein. . . . Whatever the ultimate prospects, he cannot afford to postpone dividends too long.'"

Placing World War II in perspective, Miles noted that attention to incremental dividends was a hallmark of American strategy. "In part, this sensitivity to the home front stemmed from the fear, in Marshall's words, that 'a democracy cannot fight a Seven Years War.' But coupled with this was the more positive belief that the American home front deserved to see some return on the huge investment of lives and national treasure that total war demanded."

Keeping the means of warfare proportional to the ends desired—the fifth of Miles' duties of the strategist—rose in importance as weapons became more lethal. Decisions were made not to employ nuclear weapons in the Korean and Vietnam Wars; nevertheless, the tonnages of ordnance placed on targets often exceeded those in World War II—with seemingly little contribution towards achieving strategic goals. In the latter war, however, there is much doubt as to whether any of the functions of the strategist were carried out with significant success.

This raises the question of whether the education and training of American military strategists was adequate in the decades before the Vietnam War. More significantly, it questions whether the duties of the strategist, as outlined by Colonel Miles, are being taught today to future theorists and practitioners of the strategic art.

LEARNING TO BE A STRATEGIST

Every senior commander with strategic responsibilities has a staff of field grade officers that prepares strategic concepts and develops them into plans and orders. The Army's need for a corps of "professional strategists" to man these staffs grew as America's security requirements spawned a worldwide network of American and Allied agencies. In the mid-seventies, the Army's chief of operations set out to identify career officers who had appropriate training and experience, and to tag them for future assignment in the growing strategic arena. The goal was elusive, however. Who was to define the credentials for membership in this exclusive club? Are foreign area specialists automatically "strategists?" Or military historians? Nuclear weapons experts? Political scientists? Should one have studied the Philosophers of War? Or the biographies of men who have practiced the duties of the strategist in previous wars?

Fortunately, the faculties of the nation's War Colleges (Army, Naval, Air, and National) have worried this problem for nearly as long as they have been in existence. At the end of the last century, Naval War College President Alfred Thayer Mahan produced *The Influence of Sea Power on History, 1660–1783,* whose message was taught to naval officers at Newport. In time, it revolutionized the theory and practice of naval strategy throughout the world. At the turn of the century, the Army War College was established as a strategic planning arm for the War Department General Staff. In the thirties, it harbored the handful of military strategists who developed, with their Navy counterparts, the "Color Plans" that initially provided for the nation's defense in the coming global conflict.

After World War II, each of the four colleges developed an extensive course in strategy for their students, an experienced body of officers with more than 15 years of active service. These courses were designed to increase their understanding of the strategic issues and policies with which they must cope as future senior commanders. In 1984, for example, an excellent book of readings, *Military Strategy: Theory and Practice,* provided Army War College students with a brief introduction to Clausewitz, Liddell Hart, and Sun Tzu, moving quickly to articles on the strategic problems inherent in nuclear weapons, arms control, limited and low-intensity warfare, space technology, and the ethical environment of a free society.

The best place to begin the study of strategy is in the schoolhouse, with a specially educated faculty and a structured program of reading, writing, and discussion. The War College program sharpens the minds of those going directly to strategic assignments. But it would be foolish for a commander who aspires to excellence to wait until the age of forty to begin thinking about an art that is so theoretical and complex. Fortunately, in earlier trips to the schoolhouse the soldier can obtain the basic knowledge that will put him on the path of the professional strategist. The ROTC cadet using the text *American Military History* has an excellent introduction to the meaning and importance of strategy by the editor, Maurice Matloff. In the branch schools, the officer learns of the strategic setting for his tactical operations. At the Command and General Staff College, the field grade officer can study Clausewitz in the Advanced Military Studies Program, and can enroll in history electives encompassing the strategic dimensions of American, Allied, and Soviet military power.

In the many courses in national security affairs that populate colleges and military schools, one finds extensive analysis of the background and contemporary status of the issues facing the modern strategist. The leading textbook, Jordan and Taylor's *American National Security,* describes the institutions and processes through which security policy is made. This text is an excellent reference for strategic thinkers, in chapters on nuclear strategy, limited war, low-intensity conflict, alliances, arms control, and the issues for American concern in East Asia, the Middle East, Latin America, and the Soviet Union.

History courses, like those taught by Colonel Miles on the theory and practice of designing and carrying out national strategy, deal more directly with the tasks and duties of the strategists themselves. In such courses, the following books are often used:

Brodie, *Strategy in the Missile Age*.
_____, *War and Politics*.
Clausewitz, *On War*.
Fuller, *The Conduct of War, 1789–1961*.
Gray, *Strategic Studies: A Critical Assessment*.
Greenfield, *American Strategy in World War II*.
Howard, *The Causes of War and Other Essays*.
_____, *Studies in War and Peace*.
_____, *The Theory and Practice of War*.
Liddell Hart, *Strategy*.
_____, *Thoughts on War*.
Osgood, *Limited War Revisited*.
Phillips, *The Roots of Strategy*.
Sun Tzu, *The Art of War*.
Weigley, *The American Way of War*.

Those who have not become acquainted with these books in the schoolhouse may find them unsuitable for beginning a private study program in strategic affairs. Biographies and case studies are read better along with a small reference library that can be consulted to put them in historical and theoretical perspective. A 3-volume set of references might include: *A Guide to the Study and Use of Military History*, with Jay Luvaas' chapter "The Great Military Historians and Philosophers"; Edward Mead Earle's *Makers of Modern Strategy*, a superb collection of biographical essays on strategic theorists and practitioners, from Machiavelli to Hitler; and Russell F. Weigley's *The American Way of War*, whose chapters tell chronologically the history of United States military strategy and policy from the Revolution to Vietnam.

Stephen Ambrose's *The Supreme Commander* provides a masterful portrayal of General Eisenhower's most important strategic decisions during World War II. The book opens with Eisenhower's reporting in to Chief of Staff George C. Marshall a few days after the destruction of the American fleet at Pearl Harbor. Marshall outlined the perilous position of American forces in the Philippines, and asked "What should be our line of action?'' In a few hours, Eisenhower wrote a 4-part recommendation that included building a strong base of operations in Australia for a counteroffensive. This became the core of American and Allied strategy, and Eisenhower went on to outline the global strategy that would dominate American thinking for the remainder of the war. He was one of those rare professional soldiers who was also a self-taught professional strategist. The best biography of his early preparation for this life is Kenneth Davis' *Soldier of Democracy.*

The second volume of Forrest Pogue's *George C. Marshall* is entitled *Ordeal and Hope, 1939–1942.* It tells the Army story of how a small group of statesmen and soldiers designed and implemented American grand strategy—and its military, naval, and air components—that produced victory in man's greatest global war. Marshall redefined American strategy after the fall of France in 1940. He prescribed steps for mobilizing America's industry and manpower for war against Germany and Japan. He was constant in his advice to the Commander In Chief on Anglo-American coalition strategy and the military diplomacy therein.

For a vivid portrayal of the successes and failures of earlier strategic leaders, one reads about Moltke, Jellicoe, Petain, and Ludendorff in Corelli Barnett's *The Swordbearers: Supreme Command in the First World War.* Walter Goerlitz's *The History of the German General Staff, 1657–1945* provides a succinct collective biography of an elite caste of military men who designed the strategic policy of the Prussian and German states through successive generations. For George Washington's formulation of a flexible strategy for defeating a more powerful British force, see Dave Palmer's *The Way of the Fox.* In Herold's *The Mind of Napoleon,* one finds a treatise on the many kinds of strategy, written in Bonaparte's own words.

Great strategic turning points in man's history have attracted some of the best writers. The classic example is Thucydides' account of the tragic Athenian attack on Syracuse in the Peloponnesian Wars (c. 414 B.C.). Now referred to as ''The Greek Vietnam,'' this strategic misadventure is being studied in colleges and service schools for its remarkable parallels. M.I. Finley's *The Portable Greek Historian* is one of several sources for this account.

The fall of France to Hitler's armies has provided several outstanding interpretations of a failed strategy. One is Alistair Horne's *To Lose a Battle: France, 1940,* which underlines the social and political malaise that preceded the debacle. The army's failure to respond in time of crisis is described in *Strange Defeat,* written by the great French historian, Marc Bloch, before he was executed by the Nazis.

For America's great strategic failure in 1941, Roberta Wohlstetter's *Pearl Harbor* may still be the best telling of the story. A different kind of strategic turning point occurred in Worrld War II, when the Allies penetrated the German system for encoding messages and were able thereby to tailor more closely their strategic and operational decisions. The full revelation of this drama now appears in Ronald Lewin's *Ultra Goes to War,* an excellent text on the impact of intelligence systems on strategy.

Future commanders can improve their thinking about American strategy vis-à-vis the Russians with two books. The fictional account of General Sir John Hackett, *The Third World War: The Untold Story,* forecasts the internal collapse of the Soviet government in a short conflict with the West. American officers who might command Army units in such a war have found a wealth of detail in this book on weapons, command and control systems, and strategic theories that might govern their participation in such a war. For a better understanding of the nuclear confrontation between the super powers, Lawrence Freedman provided *The Evolution of Nuclear Strategy* in 1981.

In the long run, reading by future commanders must focus on how Americans select their national goals, conduct their diplomacy, raise and maintain their forces, design their strategy, and deploy and sustain their forces in war. Weigley appropriately called this process *The American Way of War.* Americans handled this well in World War II, due partly to the political genius of the Commander in Chief. This story is told best by James McGregor Burns in *Roosevelt: Soldier of Freedom.*

The diplomatic face of American strategic history is stated no better than in the writing of George F. Kennan, whether in his short *American Diplomacy, 1900–1950;* his *Memoirs, 1925–1950;* or his many books on Russia and her relations with Europe and America. For a totally different view of the art of dealing with Russians and other adversaries, Herb Cohen's *You Can Negotiate Anything* provides unusual advice for commanders in all their roles.

Since World War II the strategic competence of America's leadership has faltered, as it has engaged increasingly in low-intensity warfare. The evidence of this ineptness was signalled in 1958 in Lederer and

Burdick's novel, *The Ugly American*. In 1966, Senator J. William Fulbright described the flawed political and strategic outlook contributing to the tragedy in Vietnam in a book entitled *The Arrogance of Power*.

In addition to Summers' *On Strategy,* two books stand out as good critiques of the strategy of the Vietnam War. Stanley Karnow's *Vietnam: A History* is probably the most comprehensive one-volume book on the war. In *The 25-Year War,* General Bruce Palmer views the conflict from the standpoint of a military participant, and takes his professional colleagues to task for failing to be more outspoken regarding the deficiencies of American strategy. The best critique of the effect of the news media on strategic goals in Vietnam is Peter Braestrup's *The Big Story*.

The effectiveness of Americans in raising, organizing, training, and commanding military forces has been the subject of three recent books. Martin van Creveld compared the performance of the German and American armies in World War II in his 1982 *Fighting Power*. He found the German Army more effective in battle because of better organization and training and superior quality in their officer corps. In *National Defense,* James Fallows decries an excessive devotion to management theory and "dilettantism" among officers in matters of strategy and operational theory. These themes are taken up again in Edward Luttwak's 1984 *The Pentagon and the Art of War*. He attributes bad strategic and operational decision-making to poor organization and a "bloated" officer corps. He would reform the Joint chiefs of Staff and create a special category of senior officers who could transcend Army, Navy, and Air Force parochialism and bias. Criticism and recommendations for change have been the lifeblood of The American Way of War since the Republic was founded.

IN SUM

Historians write that professional soldiers tend to fight the present war as they did the last one. They might also write that the makers of strategy tend to design for an era gone by rather than meet the challenges of the present.

The statesmen of the thirties were schooled to believe that their predecessors had stumbled into World War I carelessly and needlessly. (The best analysis of this unfolding of disastrous events is Barbara Tuchman's *The Guns of August*). To avoid the same debacle, they backed away from international confrontation and tried to appease the bullies in Germany, Italy, and Japan. The consequence was a global war that was almost lost.

American statesmen of the fifties and sixties learned well the lessons of World War II. They vowed never to be "soft on Communism" and confronted Red forces as Hitler should have been confronted. Now, the next generation of statesmen seems to be bent on "No More Vietnams," as if the era of the sixties will repeat itself in the nineties.

Men and women who would be "professional strategists" must be realists who recognize each international situation for what it is, not for how it will fit into the sloganeering of the time. They must be able to distinguish between real power and apparent power, between truth and propaganda, between national interest and vested interest, and between the dangers of arrogance and the dangers of weakness. Reading the history of the theory and practice of strategy can provide the knowledge and insights to help identify these distinctions. But the trick is to decide which parts of this history apply directly to the real situation at hand. This calls for experience. It also calls for the development of a personal philosophy concerning the importance of Plato's justice, temperance, wisdom, and courage in the affairs of modern man. If professional soldiers are rarely professional strategists, it is partly because of this void in their learning. Filling that void is not the subject of a book on "The Challenge of Command." Rather, it is the subject of "The Challenge of Being a Responsible Human Being."

The reminders that soldiers must become better strategists are all about us in the legacy from Vietnam. Mine is a picture of a company commander in the 101st Airborne Division, Captain Paul W. "Bud" Bucha, who in earlier years had asked the question, "What should I read?" Now he was in the jungle, kneeling by his radio-telephone operator, microphone in hand. His company was surrounded, casualties were heavy. For his bravery in this action he later received the Congressional Medal of Honor. But in this picture there were tears in his eyes. He wrote on it: "This picture was taken at the moment I realized the high price we pay for the pursuit of undefined goals."

Reading for The Strategist

In the following summary of books discussed in this chapter, those marked with an asterisk are of highest priority for the development of knowledge and insights in strategy.

* Stephen Ambrose, *The Supreme Commander: The War Years of General Dwight D. Eisenhower*.

 Correlli Barnett, *The Swordbearers: Supreme Command in The First World War*.

 Marc Bloch, *Strange Defeat*.

 Brian Bond, *Liddell Hart: A Study of His Military Thought*.

 Peter Braestrup, *Big Story*.

 Bernard Brodie, *Strategy in the Missile Age*.

* _____, *War and Politics*.

 James McGregor Burns, *Roosevelt: The Soldier of Freedom*.

 Karl von Clausewitz, *On War* (1976 Princeton Press Edition, edited by Paret and Howard).

 Herb Cohen, *You Can Negotiate Anything*.

 Martin van Creveld, *Fighting Power: German and U.S. Army Performance, 1939–1945*.

 _____, *Supplying War: Logistics from Wallenstein to Patton*.

 Kenneth S. Davis, *Soldier of Democracy: A Biography of Dwight Eisenhower*.

* Edward Mead Earle, *Makers of Modern Strategy: Military Thought from Machiavelli to Hitler*.

 James Fallows, *National Defense*.

 M.I. Finley (ed), *The Portable Greek Historian*.

 Lawrence Freedman, *The Evolution of Nuclear Strategy*.

 William Fulbright, *The Arrogance of Power*.

J.F.C. Fuller, *The Conduct of War, 1789–1961*.

Walter Goerlitz, *History of the German General Staff*.

Colin Gray, *Strategic Studies: A Critical Assessment*.

* Kent R. Greenfield, *American Strategy in World War II*.

_____ , *Command Decisions*.

General Sir John Hackett, *The Third World War: The Untold Story*.

J. Christopher Herold, *The Mind of Napoleon*.

Alistair Horne, *To Lose a Battle: France, 1940*.

Michael Howard, *The Causes of Wars and Other Essays*.

_____ , *The Mediterranean Strategy in the Second World War*.

* _____ , *Studies in War and Peace*.

_____ , *The Theory and Practice of War*.

Antoine H. Jomini, *The Art of War*.

Amos A. Jordan and William J. Taylor, *American National Security: Policy and Process*.

Stanley Karnow, *Vietnam: A History*.

* George F. Kennan, *American Diplomacy, 1900–1950*.

_____ , *Memoirs, 1925–1950*.

Walter Laquer, *Guerrilla: A Historical and Critical Study*.

* William J. Lederer and Eugene Burdick, *The Ugly American*.

* Ronald Lewin, *Ultra Goes to War*.

* B.H. Liddell Hart, *Strategy*.

_____ , *Thoughts on War*.

* Edward Luttwak, *The Pentagon and The Art of War: The Question of Military Reform*.

Alfred Thayer Mahan, *The Influence of Sea Power on History, 1660–1783*.

Maurice Matloff, *American Military History*.

Robert E. Osgood, *Limited War Revisited*.

Dave R. Palmer, *The Way of The Fox: American Strategy in The War for America, 1775–1783*.

* General Bruce Palmer, *The 25-Year War*.

Thomas R. Phillips, *Roots of Strategy: A Collection of Military Classics*.

Forrest C. Pogue, *George C. Marshall: Ordeal and Hope, 1939–1942*.

Harry G. Summers, Jr., *on Strategy: The Vietnam War in Context*.

Sun Tzu, *The Art of War*.

Barbara Tuchman, *The Guns of August*.

U.S. Army War College, *Military Strategy: Theory and Practice*.

* Russell F. Weigley, *The American Way of War: A History of United States Military Strategy and Policy*.

Roberta Wohlstetter, *Pearl Harbor: Warning and Decision*.

The Commander As Mentor

Some 250 works of biographers, historians, professional soldiers, social scientists, and philosophers have been considered in this book. Soldiers who read 4 or 5 of these works every year for 20 years, integrating this knowledge with their field experience, can follow in the steps of predecessors who distinguished themselves for excellence in military command. They can build a vision of their military selves, and enhance their capacity for the mental, moral, and action-oriented requirements of the military commander. They can become better tacticians, warriors, moral leaders, and strategists. They can undergird the courage, the sense of duty, and the dedication to justice that have distinguished soldiers of greatness throughout history. Finally, their greatest reward can come from their duty as trainers, and as mentors to the proteges who will look to them for the wisdom and skills that they seek.

My first mentor was Brig. Gen. Arthur G. Trudeau. He gave me two mind-expanding books: Balthazar Gracian's *The Art of Worldly Wisdom,* the maxims of an 18th Century Spanish Jesuit; and Peter B. Kyne's *The Go-Getter: A Story That Tells You How To Be One.* My second mentor, Colonel George A. Lincoln, thought the case of this protege was much more serious. He sent me off to Princeton for two years of reading. This process of mentoring, perhaps most celebrated in the tutoring of Dwight Eisenhower by Fox Conner, goes on today. A visitor to the office of Lt. Gen. John Galvin in Germany is apt to be given a choice between Hanson Baldwin's *Tiger Jack,* the biography of World War II commander John S. Wood, and Alistair Horne's *To Lose a Battle: France, 1940.*

At lower levels of command, some mentors are having the battalion duty officer prepare to critique a book during his 24-hour tour of duty. Others are developing terrain walks, monitoring correspondence courses, or asking officers to teach classes from books on tactics, leadership, or combat operational history. With the dramatic post-Vietnam resurgence of excellence in the corps of noncommissioned officers, senior commanders are finding it useful to redirect officer energy to professional development, while NCO's assume more responsibility for daily operations.

A case in point: In the spring of 1985, Second Lieutenant Gregory W. Cook reported to his first unit, the Third Squadron of the Second Armored Cavalry Regiment in Germany. When they had completed the Annual General Inspection, the squadron commander, Lt. Col. David Bird, packed off some 20 of the officers to Verdun, France, for a 3-day study of the World War I fortifications. The authority was Alistair Horne's *The Price of Glory: Verdun 1916.* When Cook returned, he was impressed more than ever with the importance of maneuver in warfare.

Also in 1985, a study group on officer professional development, headed by Lt. Gen. Charles W. Bagnal, urged the Officer Corps to rekindle their commitment and skills as mentors. This was to start in the schoolhouse by pairing each student with a faculty member who would monitor his progress through the various courses.

Asking the unit commander in the field to serve as mentor for his junior officers is not as easy to carry out. His role as a trainer had always been rather clear-cut—to see that his people knew their specific jobs and could perform them adequately in wartime. But if the mentor's role is to make proteges better leaders, better decision-makers, and more courageous soldiers, who is to say what the "curriculum" should be? What should proteges and mentors talk about? How many commanders are knowledgeable and skillful enough to lead such discussions?

The contention of these commentaries is that the core of the mentor's curriculum lies in these 250 books, along with others that are suitable for the environment particular to each mentor-protege situation. It is the task of the commander/mentor to select the few books that seem to fit best his scheme of things, to make the books available, and finally to allocate the time and method for individual or seminar discussion.

People who are unaccustomed to reading say that there is no time in troop units for such activity. This tends to be true when units are deployed on training missions or going through the intense upgrading of readiness standards. Under these circumstances, officer professional development meetings get postponed and mentoring sessions are diverted into discussion of the field problem at hand. But outside these cycles of intense activity, a surprising amount of officer reading and study goes on, generally on the part of officers who have a vision of themselves as constant learners, and have a reading plan.

What subject matter seems best suited for mentors and proteges at the battalion level? A discussion of professionalism, command, leadership, and management (Chapters 1 and 2 of this book) provides good background. Company-level leadership (Chapter 3), tactics (Chapter 4), and the warrior spirit (Chapter 5) are all manageable topics and germane to the unit's mission. A discussion of honesty and integrity (Chapter 6) is best suited to units with that particular problem. The more philosophical debates about the soldier's duty (Chapter 7) and military strategy (Chapter 8) are for specialized situations, particularly individual one-on-one mentor-protege relationships.

How does a mentor know if he is preparing his people for the kind of warfare they are most likely to fight? For commanders in American units in Europe, the correct focus is clearly on the AirLand Battle doctrine of medium-and high-intensity warfare. For stateside units that are deployable to Third World crises, the focus is less clear. The literature on the many varieties of low-intensity conflict is limited; some analysts doubt that the principles, doctrines, and techniques of warfare at the higher levels are useful at these lower levels. To train officers in the politics, military forces, and language of target areas may be beyond the expertise of the mentor. This is one reason why it is difficult to prepare strike force units psychologically for operating in the Middle East and Latin America. In short, some military subjects may not be suitable for the mentor.

It is the commander-as-trainer who attempts to prepare his people for missions they might embark on tomorrow morning. As mentor, however, the commander looks more towards the horizon, to the proteges' potential in years to come. His chief function is to cause his people to become better *learners*. He teaches them how to learn from reading, writing, and discussion. He motivates them to teach themselves. As such, he might be as concerned about *whether* they read as he is about *what* they read.

Recently, when Brig. Gen. Bill Stofft became the Army Chief of Military History, I asked him once again what an officer should read. He suggested that I was asking an important, but only secondary question. He argued that they must first *want* to pick up a book and read it. To do so, it must be a book on a subject of interest to them. If, after the first chapter, it no longer interests them, they can put it down and move on to the next book. But there has to be an initial movement from nonreader to reader.

The mentor must learn to distinguish between the officers who seek the right book, and the nonreaders in whom an interest must be fostered. Very often, the best approach for the latter is to recommend chapters rather than whole books, or the right articles in the better service journals, such as *Military Review, Parameters, Army,* and *Armor.* This Epilogue ends with a suggested program of "first books" that mentors can employ with proteges who have yet to become readers.

In 1978, the Chief of Staff's "Review of Education and Training for Officers" (RETO) recommended that junior officers read 8 books as part of their requirement for meeting the second level of Military Qualification Standards. These books would be chosen from lists published by the Department of the Army; battalion commanders would discuss these books with their officers, and certify to the Army personnel center that the requirement had been met. Further, it was recommended that officers read 16 additional books as part of the MQS III certification for selection for staff college and promotion to major. The men who drafted these recommendations had commanded battalions and spent many years in troop units. They believed that only through written mandates would professional development be able to compete for officer time, with the myriad of other written and verbal demands from higher headquarters.

In succeeding years, the reading lists were developed and professional reading programs were introduced into the service schools and colleges. By the spring of 1985, the essential conditions for an Army Professional Reading Program were in place. General William R. Richardson, the commander of the U.S. Army Training and Doctrine Command and long a champion of the study of military history, was seeking the best plan for providing all officers with the directives and means for pursuing a reading program throughout their careers. Again, the Combat Studies Institute at Fort Leavenworth, now headed by Colonel Dennis Frasche, was tasked to provide the historian's pespective.

In a sense, this quest for the right reading was a return to the past. General of the Army Omar N. Bradley once described how he and his lieutenant colleagues met one evening a week in the twenties and thirties to discuss their reading in tactics. "We studied everything we could get our hands on. You start working hard right from the first. You can't say later on in life I will start studying. You have got to start in the beginning." (Puryear, *Nineteen Stars,* 385)

These words were probably inspired by a mentor who saw a potential five-star general in his organization. Possibly, that mentor was steely-eyed George C. Marshall.

First Books for Officers Who Are To Become Readers

Ch. 1 Visions of Command:

E.D. Swinton, *The Defence of Duffer's Drift*.
Michael Shaara, *The Killer Angels*.

Ch. 2 The Commander:

One book from the cited biographies or memoirs of Napoleon, Lee, Grant, Jackson, Marshall, MacArthur, Patton.
Field Marshal Earl Archibald P. Wavell, *Soldiers and Soldiering*.

Ch. 3 The Company Leader:

Colonel Dandridge "Mike" Malone, *Small Unit Leadership*.
S.L.A. Marshall, *Men Against Fire*.

Ch. 4 The Tactician:

T.R. Fehrenbach, *This Kind of War*.
One book from the Germans: Rommel, Guderian, von Manstein, or von Mellenthin.

Ch. 5 The Warrior:

Lord Moran, *The Anatomy of Courage*. James R. McDonough, *Platoon Leader*.

Ch. 6 The Moral Arbiter:

Anton Myrer, *Once An Eagle*.

Ch. 7 Duty and Justice:

C.S. Forester, *The General*.
Edgar F. Puryear, Jr., *Nineteen Stars*.

Ch. 8 The Strategist:

Basil H. Liddell Hart, *Strategy,* Part IV.
Early chapters from Stephen Ambrose *Supreme Commander: The War Years of General Dwight D. Eisenhower* or Forrest C. Pogue, *George C. Marshall: Ordeal and Hope, 1939–1942*.

Bibliography

Adan, Avrahan. *On the Banks of the Suez: An Israeli General's Personal Account of the Yom Kipper War*. San Rafael, Calif: Presidio Press, 1980.

Adler, Mortimer J. *Aristotle for Everybody: Difficult Thought Made Easy*. New York: Bantam Books, 1978.

_____, *Six Great Ideas*. New York: MacMillan Publishing Co., Inc., 1981.

Albright, John; Cash, John A.; and Sandstrom, Allen E. *Seven Firefights in Vietnam*. U.S. Army Center of Military History. Washington, D.C.: Government Printing Office, 1970.

Ambrose, Stephen E. *The Supreme Commander: The War Years of General Dwight D. Eisenhower*. Garden City, N.Y.: Doubleday & Company, Inc., 1968.

Arnold, Henry H. *Global Mission*. New York: Harper, 1949.

Baldwin, Hanson W. *Tiger Jack*. Ft. Collins, Colo.: Old Army Press, 1979.

Barnett, Correlli. *The Swordbearers: Supreme Command in The First World War*. New York: Signet, 1965.

Barzun, Jacques, and Graff, Henry F. *The Modern Researcher*. Rev. Ed. New York: Harcourt, Brace & World, Inc., 1970.

Baynes, John. *Morale: A Study of Men and Courage*. New York: Frederick A. Praeger Publishers, 1967. To be reprinted by Avery Publishing Group in 1986.

Beaumont, Roger A. *Military Elites*. Indianapolis, Ind.: Bobbs-Merrill, 1974.

Bernstein, Carl, and Woodward, Robert. *All The President's Men*. New York: Simon and Schuster, 1974.

Blanchard, Kenneth; Zigarmi, Patricia; and Zigarmi, Drea. *Leadership and The One Minute Manager*. New York: William Morrow, 1985.

Bloch, Marc. *Strange Defeat*. New York: W.W. Norton and Co., 1968.

Blumenson, Martin, *The Patton Papers*. Vol. I: *1881–1940*. Vol. II: *1940 –1945*. Boston: Houghton Mifflin Co., 1972 and 1974.

Blumenson, Martin, and Stokesbury, James L. *Masters of the Art of Command*. Boston: Houghton Mifflin Co., 1975. To be reprinted by Avery Publishing Group, 1986.

Bok, Sissela. *Lying: Moral Choice in Public and Private Life*. New York: Random House, Inc., 1978.

_____. *Secrets: On The Ethics of concealment and Revelation*. New York: Vintage Books, 1984.

Bond, Brian. *Liddell Hart: A Study of His Military Thought*. London: Cassell, 1977.

Bradley, Omar N. *A Soldier's Story*. New York: Henry Holt and Co., 1951.

Braestrup, Peter. *Big Story: How the American Press and Television Reported and Interpreted the Crisis of Tet 1968 in Vietnam and Washington*. New Haven, Conn.: Yale University Press, 1983.

Brodie, Bernard. *Strategy in the Missile Age*. Princeton, N.J.: Princeton University Press, 1959.

_____. *War and Politics*. New York: The MacMillan Company, 1973.

Buck, James H., and Korb, Larry J. (ed.) *Military Leadership*. Beverly Hills, California.: Sage Publications, 1981.

Buell, Thomas B. *The Quiet Warrior: A Biography of Admiral Raymond A. Spruance*. Boston: Little Brown, 1974.

Bunting, Josiah. *The Lionheads*. New York: George Braziller, 1972.

Burns, James MacGregor. *Leadership*. New York: Harper and Row, 1978.

_____. *Roosevelt: The Soldier of Freedom*. New York: Harcourt, Brace, Jovanovitch. 1970.

Caesar. *The Conquest of Gaul*. Trans. by S.A. Handford. Harmondsworth, England: Penguin, 1982.

Caputo, Phillip. *A Rumor of War*. New York: Ballantine Books, 1977.

Chandler, David G. *The Campaigns of Napoleon*. New York: The MacMillan Co., 1966.

Charol, Michael. *The Mongol Empire: It's Rise and Legacy*. New York: The MacMillan Co., 1940.

Clarke, Gen. Bruce C. *Guidelines for the Leader and the Commander*. Harrisburg, Pa.: Stackpole Books, 1963.

Clausewitz, Carl von. *On War*. Ed. and trans. by Michael Howard and Peter Paret. Princeton, N.J.: Princeton University Press, 1976.

Coffman, Edward M. *The War To End All Wars: The American Military Experience In World War I*. New York: Oxford University Press, 1968.

Cohen, Herb. *You Can Negotiate Anything*. Secaucus, N.J.: Lyle Stuart, Inc., 1980.

Collins, Lt. Gen. Arthur S. Jr. *Common Sense Training: A Working Philosophy for Leaders*. Novato, Calif.: Presidio Press, 1978.

Collins, Gen. J. Lawton. *Lightning Joe: An Autobiography*. Baton Rouge, La.: Louisiana State University Press, 1979.

Connell, Evan S. *Son of the Morning Star*. San Francisco, Calif.: North Point Press, 1984.

Cozzens, James G. *Guard of Honor*. New York: Harcourt, Brace, and Co., 1948.

Crane, Stephen. *The Red Badge of Courage*. New York: D. Appleton and Co., 1926.

Creveld, Martin van. *Command in War*. Cambridge, Mass.: Harvard University Press, 1985.

_____. *Fighting Power: German and U.S. Army Performance, 1939–1945*. Westport, Conn.: Greenwood Press, 1982.

_____. *Supplying War: Logistics from Wallenstein to Patton*. Cambridge, England: Cambridge University Press, 1977.

Crocker, Lawrence P. *The Army Officer's Guide*. 42nd ed. Harrisburg, Pa.: Stackpole Books, 1981.

Davis, Kenneth S. *Soldier of Democracy: A Biography of Dwight Eisenhower*. Garden City, N.Y.: Doubleday, Doran & Co., 1945.

Doughty, Robert A. *The Evolution of U.S. Army Tactical Doctrine, 1946–1976*. A Leavenworth Paper. Combat Studies Institute. Fort Leavenworth, Kan.: U.S. Army Command and General Staff College, 1979.

Dyke, Arthur J. *On Human Care: An Introduction to Ethics*. Nashville, Tenn.: Abingdon Press, 1977.

Earle, Edward Mead (ed.). *Makers of Modern Strategy: Military Thought from Machiavelli to Hitler*. Princeton, N.J.: Princeton University Press, 1944.

Eisenhower, Dwight D. *At Ease: Stories I tell My Friends*. New York: Avon Books, 1967.

_____. *The Eisenhower Diaries*. Ed. by Robert H. Farrell. New York: W.W. Norton & Co., 1981.

Eisenhower, John S.D. *The Bitter Woods*. New York: G.P. Putnam's Sons, 1969.

English, John A. *A Perspective on Infantry*. New York: Praeger, 1981.

Epictetus. *The Discourses*. Vol. 12 of *Great Books of the Western World*. Chicago: Encyclopaedia Britannica, Inc., 1952.

Fallows, James. *National Defense*. New York: Vintage Books, 1981.

Farr, Charles. *From the Jaws of Victory*. New York: Simon and Schuster, 1971.

Fehrenbach, T.R. *This Kind of War: A Study in Unpreparedness*. New York: The MacMillan Co., 1963.

Finley, M.I. (ed.) *The Portable Greek Historians: The Essense of Herodotus, Thucydides, Xenophon, Polybius*. New York: Penguin Books, 1977.

Ford, Daniel. *Incident at Muc Wa*. Garden City, N.Y.: Doubleday & Co., Inc., 1967.

Forester, Cecil S. *Beat to Quarters*. New York: Little, Brown & Co., 1937.

_____. *The General*. Annapolis, Md.: The Nautical and Aviation Publishing Co., 1982.

_____. *Rifleman Dodd*. Also published as *Death to the French*. Salem, N.H.: Merrimack Publishing Circle, 1978.

Fox, Robin L. *Alexander the Great*. New York: Dial Press, 1974.

Frankel, Nat, and Smith, Larry. *Patton's Best: An Informal History of the 4th Armored Division*. New York: Hawthorne, 1978.

Frankena, William K. *Ethics*. Englewood Cliffs, N.J.: Prentice-Hall, Inc., 1973.

Frederick the Great on The Art of War. Translated and edited by Jay Luvaas. New York: The Free Press, 1966.

Freedman, Lawrence. *The Evolution of Nuclear Strategy*. New York: St. Martin's Press, 1981.

Freeman, Douglas S. *R.E. Lee: A Biography*. 4 Vols. New York: Charles Scrbner's Sons, 1934–35. (Also abridged in 1 volume *Lee,* ed. by Richard Howell.)

Fromm, Erich. *Man for Himself*. Greenwich, Conn.: Fawcett Publications, Inc., 1947.

Fulbright, Senator J. William. *The Arrogance of Power*. New York: Vintage Books, 1966.

Fuller, J.F.C. *Generalship: Its Diseases and Their Cures. The Personal Factor in Command*. Harrisburg, Pa.: The Military Service Publishing Co., 1936.

_____. *The Conduct of War, 1789–1961*. New York: Funk and Wagnals Publishing Co., Inc., 1968.

_____. *The Generalship of Alexander the Grat*. New Brunswick, N.J.: Rutgers University Press, 1968.

_____. *The Generalship of Ulysses S. Grant*. Bloomington, Ind.: Indiana University Press, 1958.

_____. *Grant and Lee: A Study in Personality and Generalship*. Bloomington, Ind.: Indiana University Press, 1957.

_____. *Julius Caesar: Man, Soldier, and Tyrant*. New Brunswick, N.J.: Rutgers University Press, 1965.

_____. *A Short History of the Second World War*. London: Faber & Faber, 1950.

Gabriel, Richard A. and Savage, Paul L. *Crisis in Command: Mismanagement in the Army*. New York: Hill and Wang, 1978.

Galvin, John A. *Air Assault: The Development of Airmobile Warfare*. New York: Hawthorne, 1969.

Glantz, David M. *August Storm*. 2 vols. 1. *Soviet Tactical and Operational Combat in Manchuria, 1945*. 2. *The Soviet 1945 Strategic Offensive in Manchuria*. Leavenworth Papers. Combat Studies Institute. Fort Leavenworth, Kan.: U.S. Army Command and General Staff College, 1983.

Goerlitz, Walter. *History of the German General Staff, 1657–1945*. New York: Frederick A. Praeger, 1953.

Gracian, Balthasar. *The Art of Worldly Wisdom*. New York: The MacMillan Company, 1946.

Grant, Michael. *Julius Caesar*. New York: McGraw Hill, 1969.

Grant, Ulysses S. *Personal Memoirs of U.S. Grant*. 2 Vols. New York: Charles L. Webster & Co., 1885. An abridged *Personal Memoirs of U.S. Grant* is edited by E.B. Long. New York: DeCapo, 1984.

Graves, Robert. *Goodbye to All That*. London: Jonathan Cape, 1929.

Gray, Collin. *Strategic Studies, A Critical Assessment*. Westport, Conn.: Greenwood Press, 1982.

Gray, J. Glenn. *The Warriors*. New York: Harper and Row, Publishers, 1967.

The Great Ideas. A Syntopicon of Great Books of the Western World. 2 vols. Edited by Mortimer C. Adler and the Faculty of the University of Chicago. Chicago: Encyclopaedia Britannica, Inc., 1952.

Greenfield, Kent R. *American Strategy in World War II: A Reconsideration*. Malabar, Fla.: Robert E. Krieger Publishing Co., 1983.

————. (ed.) *Command Decisions*. U.S. Army Chief of Military History. Washington, D.C.: Government Printing Office, 1984.

Griess, Thomas E. (ed.) *The West Point Military History Series*. 15 vols. Wayne, N.J.: Avery Publishing Group, 1985–86.

Guderian, Heinz. *Panzer Leader*. 1952. Reprint, New York: Ballantine Books, 1980.

Gugeler, Russell. *Combat Actions in Korea*. U.S. Army Center of Military History. Washington, D.C.: Government Printing Office, 1970.

Hackett, Sir John. *The Third World War: The Untold Story*. New York: The MacMillan Publishing Co., 1982.

Haines, William W. *Command Decision: A Play*. New York: Random House, 1948.

Hamilton, Nigel. *Monty*. 2 vols. 1. *The Making of a General, 1887–1942*. 2. *Master of the Battlefield, 1942–1944*. New York: McGraw-Hill, 1981–1984.

Hays, Samuel H. and Thomas, William H. *Taking Command: The Art and Science of Military Leadership*. Harrisburg, Pa.: Stackpole, 1967.

Heinl, Robert D., Jr. *Dictionary of Military and Naval Quotations*. Annapolis, Md.: U.S. Naval Institute, 1966.

Heller, Charles E. and Stofft, William A. (ed.) *American First Battles*. Lawrence, Kan.: The Kansas University Press, to be published in winter, 1985.

Henderson, W. Darryl. *Cohesion: The Human Element in Combat*. Washington, D.C.: National Defense University Press, 1985. Available from GPO.

Henderson, Colonel G.F.R. *Stonewall Jackson and the American Civil War*. 2d ed. New York: Longmans, Green & Co., 1936. 1st ed. in 2 vols, London, 1898.

Herold, J. Christopher. *The Mind of Napoleon: A Selection of His Written and Spoken Words*. New York: Columbia University Press, 1955.

Hersey, Paul and Blanchard, Kenneth H. *Management of Organizational Behavior: Utilizing Human Resources*. Englewood Cliffs, Calif.: Prentice-Hall, Inc., 1977.

Hersh, Seymour M. *My Lai 4: A Report on the Massacre and Its Aftermath*. New York: Random House, 1970.

Herzog, Chaim. *The Arab-Israeli Wars*. New York: Vintage Books, 1982.

Horgan, Paul. *A Distant Trumpet*. Greenwich, Conn.: Fawcett Publications, Inc., 1961.

Horne, Alistair. *The Price of Glory: Verdun 1916*. New York: St. Martin's Press, 1963.

_____. *To Lose a Battle: France, 1940*. New York: Penguin Books, 1979.

Hospers, John. *Human Conduct: Problems of Ethics*. New York: Harcourt, Brace, Jovanovich, Inc., 1972.

House, Jonathan M. *Towards Combined Arms Warfare: A Survey of Tactics, Doctrine, and Organization in the Twentieth Century*. Research Study 2. Combat Studies Institute. Fort Leavenworth, Kan.: U.S. Army Command and General Staff College, 1984.

Howard, Michael. *The Causes of War and Other Essays*. Cambridge, Mass.: Harvard University Press, 1983.

_____. (ed.) *The Mediterranean Strategy in the Second World War*. London: Weidenfeld & Nicholson, 1966.

_____. (ed.) *Restraints on War: Studies in the Limitation of Armed Conflict*. New York: Oxford University Press, 1979.

_____. *Studies in War and Peace*. New York: The Viking Press, 1970.

_____. (ed.) *The Theory and Practice of War*. London: Cassell, 1965.

_____. *War in European History*. New York: Oxford University Press, 1976.

Huntington, Samuel P. *The Soldier and the State*. New York: The Belknap Press of Harvard University Press, 1957.

Ingraham, Larry H. *The Boys in the Barracks*. Philadelphia: Institute for the Study of Human Issues, 1984.

Jacobs, Bruce. *Heroes of the Army: The Medal of Honor and Its Winners*. New York: W.W. Norton & Co., 1956.

Janowitz, Morris. (ed.) *The New Military: Changing Patterns of Organization*. New York: Russell Sage Foundation, 1964.

_____. *The Professional Soldier: A Social and Political Portrait*. Glencoe, Ill.: The Free Press, 1960.

Jessup, John E. and Coakley, Robert W. *A Guide to the Study and Use of Military History*. U.S. Army Center of Military History. Washington, D.C.: Government Printing Office, 1979.

Jomini, Antoine H. *The Art of War*. 1838. Reprint, Westport, Conn.: Greenwood Press, 1971.

Jordan, Amos A. and Taylor, William J., Jr. *American National Security*. Baltimore, Md.: The Johns Hopkins University Press, 1984.

Karnow, Stanley. *Vietnam: A History*. New York: Penguin Books, 1984.

Karsten, Peter. *Law, Soldiers, and Combat*. Westport, Conn.: Greenwood Press, 1978.

Kellett, Anthony. *Combat Motivation: The Behavior of Soldiers in Battle*. Boston: Klumer-Nijhoff Publications, 1980.

Kemble, C. Robert. *The Image of the Army Officer in America: Background for Current Views*. Westport, Conn.: Greenwood Press, 1973.

Kennan, George F. *American Diplomacy, 1900–1950*. Chicago: University of Chicago, 1984.

_____. *Memoirs, 1925–1950*. Boston: Little, Brown, 1967.

Kennedy, John F. *Profiles in Courage*. New York: Pocket Books, Inc., 1957.

Kinnard, Douglas. *The War Managers*. Wayne, N.J.: Avery Publishing Group, 1985.

Kohlberg, Lawrence. *The Philosophy of Moral Development: Moral Stages and the Idea of Justice*. New York: Harper and Row, Publishers, 1981.

Koyen, Kenneth. *The Fourth Armored Division: From the Beach to Bavaria*. Waterloo, N.Y.: Hungerford-Holbrook Co., 1949.

Kuwahara, Yasuo, and Allred, Gordon T. *Kamakaze*. New York: Ballantine Books, 1982.

Kyne, Peter B. *The Go-Getter: The Story That Tells You How To Be One*. New York: Rinehart & Company, 1921.

Laquer, Walter. *Guerrilla: A Historical and Critical Study*. Boston: Little, Brown, 1976.

Larteguy, Jean. *The Centurions*. New York: E.P. Dutton & Co., 1962.

Lederer, William J. and Burdick, Eugene. *The Ugly American*. New York: W.W. Norton & Co., 1958.

Lewin, Ronald. *Rommel as Military Commander*. Princeton, N.J.: D. Van Nostrand Co., Inc., 1968.

_____. *Ultra Goes to War*. New York: Pocket Books, 1978.

Lewis, C.S. *The Abolition of Man*. New York: MacMillan Publishing Co. Inc., 1978.

Lewy, Guenter. *America in Vietnam*. New York: Oxford University Press, 1978.

Liddell Hart, Basil H. *The Remaking of Modern Armies*. Boston: Little, Brown, 1928.

_____. *The Rommel Papers*. 1953. Reprint, New York: DeCapo, 1982.

_____. *Strategy*. 2d rev. ed. New York: New American Library, 1974.

_____. *Thoughts on War*. London: Faber & Faber, 1944. To be reprinted by Avery Publishing Group, 1986.

Longford, Elizabeth. *Wellington: The Years of the Sword*. New York: Harper and Row, 1969.

Lovell, John P. *Neither Athens Nor Sparta: The American Service Academies in Transition*. Bloomington, Ind.: Indiana University Press, 1979.

Lupfer, Timothy T. *The Dynamics of Doctrine: The Change in German Tactical Doctrine During the First World War*. Leavenworth Paper 4. Combat Studies Institute. Fort Leavenworth, Kan.: U.S. Army Command and General Staff College, 1981.

Luttwak, Edward N. *The Pentagon and the Art of War. The Question of Military Reform*. New York: Simon and Schuster Publications, 1984.

MacDonald, Charles B. *The Battle of the Huertgen Forest*. Philadelphia: Lippincott, 1963.

_____. *Company Commander*. New York: Bantam Books, 1978.

MacDonald, Charles B. and Mathews, Sidney T. *Three Battles: Arnaville, Altuzzo, and Schmidt*. U.S. Army Center of Military History. Washington, D.C.: Government Printing Office, 1952.

MacGregor, Morris J., Jr. *Integration of the Armed Forces, 1940–1965*. U.S. Army Center of Military History. Washington, D.C.: Government Printing Office, 1981.

MacIntyre, Alasdair. *After Virtue: A Study in Moral Theory*. Notre Dame, Ind.: University of Notre Dame Press, 1981.

Magnus, Phillip. *Kitchener: Portrait of an Imperialist*. New York: E.P. Dutton & Co., Inc., 1968.

Mahan, Alfred Thayer. *The Influence of Sea Power on History, 1660–1783*. 1890. Reprint, New York: Hill and Wang, 1957.

Malone, Colonel Dandridge M. *Small Unit Leadership: A Commonsense Approach*. Novato, Calif.: Presidio Press, 1983.

Manchester, William. *American Caesar: Douglas MacArthur, 1880–1964*. Boston: Little, Brown, 1978.

Manstein, Eric von. *Lost Victories*. Chicago: Henry Regnery Company, 1958.

Marshall, S.L.A. *Men Against Fire: The Problem of Battle Command in a Future War*. 1947. Reprint, Gloucester, Mass.: Peter Smith, 1978.

_____. *The River and the Gauntlet*. New York: William Morrow and Co., 1953.

Maslow, A.H. *Motivation and Personality*. New York: Harper, 1954.

Masters, John. *Bugles and a Tiger*. New York: Viking Press, 1956. To be reprinted by Avery Publishing Group in 1986.

_____. *High Command*. New York: William Morrow and Company, 1983.

_____. *The Road Past Mandalay: A Personal Narrative*. New York: Harper, 1961.

Matloff, Maurice. (ed.) *American Military History*. rev. ed. U.S. Army

Center of Military History. Washington, D.C.: Government Printing Office, 1973.

McConnell, Malcolm. *Into the Mouth of the Cat: The Story of Lance Sijan, Hero of Vietnam.* New York: W.W. Norton & Company, 1985.

McDonough, James R. *Platoon Leader.* Novato, Calif.: Presidio Press, 1985.

McFeely, William S. *Grant: A Biography.* New York: W.W. Norton & Co., 1982.

Mellenthin, F.W. von. *Panzer Battles: A Study of the Employment of Armor in the Second World War.* 1956. Reprint, New York: Ballantine Books, 1980.

Melville, Herman. *Billy Budd, Sailor: An Inside Narrative.* Indianapolis, Ind.: Bobbs-Merrill, 1975.

Messenger, Charles. *The Blitzkrieg Story.* New York: Charles Scribner's Sons, 1976.

Montgomery, Bernard Law. *The Path to Leadership.* New York: G.P. Putnam's Sons, 1961.

Moran, Lord Charles Wilson. *The Anatomy of Courage.* 1945. Reprint, London: Constable and Company, Ltd., 1966. To be reprinted by Avery Publishing Group in 1985.

Mosely, Leonard. *Marshall: Hero for Our Times.* New York: Hearst, 1982.

Moskos, Charles C. Jr. *The American Enlisted Man: The Rank and File in Today's Military.* New York: Russell Sage Foundation, 1970.

Musashi, Miyamoto. *A Book of Five Rings.* Woodstock, N.Y.: The Overlook Press, 1982.

Myrer, Anton. *Once an Eagle.* New York: Dell Publishing Company, 1970.

Naisbitt, John. *Megatrends: Ten New Directions Transforming Our Lives.* New York: Warner books, Inc., 1982.

Newman, Maj. Gen. Aubrey. *Fcllow Me: The Human Element in Leadership.* Novato, Calif.: Presidio Press, 1981.

Osgood, Robert E. *Limited War Revisited.* Boulder, Colo.: Westview Press, 1979.

Palmer, David R. *Summons of the Trumpet: A History of the Vietnam War from a Military Man's Perspective.* 1978. Reprint, New York: Ballantine Books, 1984.

————. *The Way of the Fox: American Strategy in War for America, 1775–1783.* Westport, Conn.: Greenwood Press, 1975.

Palmer, General Bruce. *The 25-Year War.* Lexington, Ken.: The University of Kentucky Press, 1984.

Parotkin, Ivan. (ed.) *The Battle of Kursk.* Moscow: Progress Publications, 1974. To be reprinted by Avery Publishing Group, 1986.

Patton, George S., Jr. *War As I Knew It.* 1947. Reprint, New York: Bantam Books, 1981.

Peters, Thomas, Jr. and Austin, Nancy. *A Passion for Excellence*. New York: Random House, 1985.

Phillips, Thomas R. (ed.) *The Roots of Strategy: A Collection of Military Classics*. 1955. Reprint, Harrisburg, Pa.: Stackpole, 1985.

Picq, Ardant du. *Battle Studies: Ancient and Modern Battle*. 1921. Reprint, Harrisburg, Pa.: The Military Service Publishing Company, 1958.

Pogue, Forrest C. *George C. Marshall*. 2 vols. I. *Education of a General, 1880 –1939*. II. *Ordeal and Hope, 1939–1942*. New York: The Viking Press, 1963 and 1966.

Puryear, Edgar F., Jr. *Nineteen Stars*. 2d ed. Novato, Calif.: Presidio Press, 1984.

Rawls, John. *A Theory of Justice*. Cambridge, Mass.: The Belknap Press of the Harvard University Press, 1971.

Renault, Mary. *The Nature of Alexander*. New York: Pantheon Books, 1975.

Ridgway, Matthew B. *Soldier: The Memoirs of Matthew B. Ridgway*. 1956. Reprint, Westport, Conn.: Greenwood Press, 1974.

Rogan, Helen. *Mixed Company: Women in the Modern Army*. New York: Putnam, 1981.

Romjue, John L. *From Active Defense to AirLand Battle: The Development of Army Doctrine, 1973–1982*. Fort Monroe, Vir.: U.S. Army Training and Doctrine Command, 1984.

Rommel, Erwin. *Attacks*. 1937. Reprint, Vienna, Va.: Athena Press, 1979.

Rustow, Dankwart A. *Philosophers and Kings: Studies in Leadership*. New York: George Braziller, 1970.

Ryan, Cornelius. *A Bridge Too Far*. New York: Simon and Schuster, 1974.

Sajer, Guy. *The Forgotten Soldier*. Trans. by Lily Emmet. New York: Harper & Row., 1971.

Sarkesian, Sam C. *Beyond the Battlefield: The New Military Professionalism*. New York: Pergamon Press, 1981.

―――――, *The Professional Army Officer in a Changing Society*. Chicago: Nelson-Hall Publishers, 1975.

Sassoon, Siegfried. *Memoirs of an Infantry Officer*. London: Faber and Faber, Ltd., 1930.

Saxe, Maurice Compte de. *Reveries or Memoirs Upon the Art of War*. 1757. Reprint, Westport, Conn.: Greenwood Press, 1971.

Schell, Captain Adolph von. *Battle Leadership*. 1933. Reprint, Quantico, Vir.: Marine Corps Association, 1982.

Scott, William G. and Hart, David K. *Organizational America*. Boston: Houghton Mifflin Co., 1979.

Semmes, Harry H. *Portrait of Patton*. New York: Paperback Library, 1970.

Shaara, Michael. *The Killer Angels*. New York: Ballantine Books, 1974.

Shakespeare, William. *The Life of Henry V*. New York: Signet Books of The New American Library, 1965.

Shazly, Lieut. Gen. Saad el. *The Crossing of the Suez.* San Francisco, Calif.: American Mideast Research, 1980.

Sherman, William T. *The Memoirs of William T. Sherman.* 1875. Reprint, Bloomington, Ind.: Indiana University Press, 1957.

Simpson, Charles M. III. *Inside the Green Berets: The First Thirty Years.* A History of the U.S. Army Special Forces. New York: Berkeley Books, 1984.

Slim, Field Marshal Sir William. *Defeat Into Victory.* New York: David McKay Co., Inc., 1961.

————, *Unofficial History.* New York: David McKay Co., 1962.

Stallworthy, Jon. (ed.) *The Oxford Book of War Poetry.* New York: Oxford University Press, 1984.

Starry, Donn A. *Mounted Combat in Vietnam.* U.S. Army Center of Military History. Washington, D.C.: Government Printing Office, 1978.

Steers, Richard M. and Porter, Lyman W. *Motivation and Work Behavior.* New York: McGraw-Hill Co., 1979.

Stogdill, Ralph M. *Handbook of Leadership: A Survey of Theory and Research.* New York: The Free Press, 1981.

Stromberg, Peter L. *A Long War's Writing: American Novels About the Fighting in Vietnam Written While Americans Fought.* Ann Arbor, Mich.: University Microfilms, 1974.

Summers, Harry G., Jr. *On Strategy: A Critical Analysis of the Vietnam War.* New York: Dell, 1982.

Sun Tzu. *The Art of War.* New York: Oxford University Press, 1963.

Swinton, Ernest D. *The Defence of Duffer's Drift.* Reprint from United Services Magazine, London: W. Clowes & Sons, Ltd., 1929. Reprint, Ft. Leavenworth, Kan.: U.S. Army Command and General Staff College, 1977. To be reprinted by Avery Publishing Group in 1985.

————, *The Study of War.* Oxford, Eng.: The Clarendon Press, 1926.

Taylor, Robert L. and Rosenbach, William E. (ed.) *Military Leadership: In Pursuit of Excellence.* Boulder, Colo.: Westview Press, 1984.

Thompson, James M. *Napoleon Bonaparte.* New York: Oxford University Press, 1952.

Thucydides. *History of the Peloponnesian War.* 5th Century B.C. Trans. by Rex Warner. Harmondsworth, Eng.: Penguin, 1972.

Tuchman, Barbara W. *The Guns of August.* 1962. Reprint, New York: Bantam Books, 1980.

————, *Practicing History.* New York: Alfred A. Knopf, 1981.

————, *Stilwell and the American Experience In China, 1911–1945.* New York: Bantam Books, 1972.

U.S. Army Center of Military History. *Style Guide.* Washington, D.C.: Government Printing Office, 1983.

U.S. Army Combat Studies Institute. *War and Doctrine.* Document P162. Fort

Leavenworth, Kan.: U.S. Army Command and General Staff College, 1983.

U.S. Army Infantry School. *Infantry in Battle*. Washington, D.C.: The Infantry Journal, 1939. Reprint, Fort Leavenworth, Kan.: U.S. Army Command and General Staff College, 1982.

U.S. Army War College. *Military Strategy: Theory and Application*. A Reference Text, 1983–1984. Carlisle Barracks, Penn.: U.S. Army War College, 1983.

U.S. Department of the Army. *Military Leadership*. Field Manual 22-100. Washington, D.C.: Government Printing Office, 1982.

U.S. Department of the Army. *Operations*. Field Manual 100-5. Washington, D.C.: Government Printing Office, 1983.

U.S. Department of the Army. *Soviet Army Operations*. Washington, D.C.: Government Printing Office, 1983.

U.S. Department of Defense. *The Armed Forces Officer*. DOD GEN-36. Written by Brig. Gen. S.L.A. Marshall, Washington, D.C.: Government Printing Office, 1975.

U.S. Military Academy. Associates, the Department of Behavioral Sciences and Leadership. *Leadership in Organizations*. West Point, N.Y.: U.S. Military Academy, 1983.

U.S. Military Academy. Associates in Military Leadership. *A Study of Organizational Leadership*. West Point, N.Y.: U.S. Military Academy, 1975.

Wakin, Malham M. *War, Morality, and the Military Profession*. Boulder, Colo.: Westview Press, 1979.

Walzer, Michael. *Just and Unjust Wars*. New York: Basic Books, Inc., 1977.

———, *Obligations: Essays on Disobedience, War, and Citizenship*. Cambridge, Mass.: Harvard University Press, 1970.

Waugh, Evelyn. *Officers and Gentlemen*. Boston: Little, Brown & Co., 1955.

Wavell, Field Marshal Earl Archibald P. *Soldiers and Soldiering*. Includes 1939 Lee Knowles Lectures *Generals and Generalship*. London: Jonathan Cape, 1953. To be reprinted by Avery Publishing Group in 1986.

Weigley, Russell F. *The American Way of War: A History of United States Military Strategy and Policy*. New York: MacMillan Publishing Co., Inc., 1973.

Williams, T. Harry. *Lincoln and His Generals*. New York: Vintage Books, 1952.

Wise, David. *The Politics of Lying: Government Deception, Secrecy, and Power*. New York: Random House, 1973.

Wohlstetter, Roberta. *Pearl Harbor: Warning and Decision*. Stanford, Calif.: Stanford University Press, 1962.

Woodham-Smith, Cecil. *The Reason Why*. New York: McGraw-Hill Book Company, Inc, 1953.

Wouk, Herman. *The Caine Mutiny*. Garden City, N.Y.: Doubleday, 1951.

About the Author

Roger Hurless Nye was commissioned in the Armor Branch of the United States Army in 1946, and served in armor and infantry units in Europe and the Korean War. He received a Master's degree from Princeton University, and graduated from the U.S. Army Command and General Staff college. He obtained his Doctoral degree in American History from Columbia University. After retiring from his position as U.S. Military Academy Professor of History in 1975, he was recalled to active duty to serve on the Army Chief of Staff's Review of Education and Training for Officers. He has participated frequently in Army policy studies on officer professional development.

Index

measured by commanders, 93
warrior's use of, 90
Visions of Our Military Selves, 1
battalion command, 20
commissioning systems and,
4
and a military future, 1
problem-solving culture
and, 4
the tactician, 72
uses of, 3–4
varieties of, 5, 16

Wakin, Col. Malham N., 15
Walzer, Michael, 120
War
attractions of, 92
causes, 144–145
conditions of just, 87
Convention, The, 87
crimes as inefficient, 89
Warrior
as barbarian, 79
British view, 80
and cohesive units, 85
four challenges to the, 90–91
Gray, J. G. on the, 92
and Halsey, 83
and McDonough, J. R., 90
and Mongols, 80
Patton, G. S. on the, 79–80
qualities of, 80

and samurai, 80
spirit, 79
Washington, Gen. George,
7, 10
as strategist, 142
Wass de Czege, Col. Huba, 74
Watergate, 112
Wavell, Field Marshal Sir
Archibald P.
on generalship, 33
on justice, 123
Weigley, Prof. Russell F., 141
Wellington, Arthur
Wellesley, Duke of,
6, 9, 87
Westmoreland, Gen.
William C., 29
Winton, Lt. Col. Harold R., 74
Wise, David, 112
Women, in Officer Corps, 5
Woodham-Smith, Cecil, 129
World War II
blitzkreig, 87
Remagen Bridge, 70
strategists of, 133–134,
137–138
and warrior spirit, 92
and weapons firing, 63
Wouk, Herman, 25
Writing
reference shelf for, 131
for self-development, 129

Xenophon, 130